Edward L. Wolf and Manasa Medikonda

**Understanding the
Nanotechnology Revolution**

Related Titles

Wolf, E. L.

Quantum Nanoelectronics

An Introduction to Electronic Nanotechnology and Quantum Computing

2009
ISBN: 978-3-527-40749-1

Wolf, E. L.

Nanophysics and Nanotechnology

An Introduction to Modern Concepts in Nanoscience

2006
ISBN: 978-3-527-40651-7

Rubahn, H.-G.

Basics of Nanotechnology

2008
ISBN: 978-3-527-40800-9

Edwards, S. A.

The Nanotech Pioneers

Where Are They Taking Us?

2006
ISBN: 978-3-527-31290-0

Fayngold, M.

Special Relativity and How It Works

2008
ISBN: 978-3-527-40607-4

Booker, R. D., Boysen, E.

Nanotechnology for Dummies

2006
ISBN: 978-0-7645-8368-1

Edward L. Wolf and Manasa Medikonda

Understanding the Nanotechnology Revolution

WILEY-VCH Verlag GmbH & Co. KGaA

The Authors

Prof. Edward L. Wolf
New York University
Polytechnic Institute
Brooklyn, New York
USA
ewolf@poly.edu

Manasa Medikonda
State University of New York at Albany
School of Nanoscale Science and Engineering
Albany, New York
USA

Cover
Nanotube by Geoffrey Hutchison, Pittsburgh, USA

All books published by **Wiley-VCH** are carefully produced. Nevertheless, authors, editors, and publisher do not warrant the information contained in these books, including this book, to be free of errors. Readers are advised to keep in mind that statements, data, illustrations, procedural details or other items may inadvertently be inaccurate.

Library of Congress Card No.: applied for

British Library Cataloguing-in-Publication Data
A catalogue record for this book is available from the British Library.

Bibliographic information published by the Deutsche Nationalbibliothek
The Deutsche Nationalbibliothek lists this publication in the Deutsche Nationalbibliografie; detailed bibliographic data are available on the Internet at <http://dnb.d-nb.de>.

© 2012 Wiley-VCH Verlag & Co. KGaA, Boschstr. 12, 69469 Weinheim, Germany

All rights reserved (including those of translation into other languages). No part of this book may be reproduced in any form – by photoprinting, microfilm, or any other means – nor transmitted or translated into a machine language without written permission from the publishers. Registered names, trademarks, etc. used in this book, even when not specifically marked as such, are not to be considered unprotected by law.

Cover   Grafik-Design Schulz, Fußgönheim
Typesetting   Toppan Best-set Premedia Limited, Hong Kong
Printing and Binding   Markono Print Media Pte Ltd, Singapore

Print ISBN: 978-3-527-41109-2

Contents

Preface *IX*

1 Discovery, Invention, and Science in Human Progress *1*
1.1 Origins of Technology, the Need for Human Survival *1*
1.2 The Industrial Revolution: Watt's Steam Engine, Thermodynamics, Energy Sources *2*
1.3 A Short History of Time: Navigation, Longitudes, Clocks *4*
1.4 The Information Revolution: Abacus to Computer Chips and Fiber Optics *5*
1.5 Overlap and Accelerating Cascade of Technologies: GPS, Nuclear Submarines *6*
1.6 Silicon and Biotechnologies: Carbon Dating, Artificial Intelligence *7*
1.7 Nanotechnology: A Leading Edge of Technological Advance, a Bridge to the Future *13*
1.8 How to Use This Book *15*
References *16*

2 Smaller Is More, Usually Better, and Sometimes Entirely New! *17*
2.1 Nanometers, Micrometers, Millimeters – Visualizing a Nanometer *18*
2.2 Moore's Law: from 30 Transistors to a Billion Transistors on One Chip and Cloud Computing *19*
2.3 Miniaturization: Esaki's Tunneling Diode, 1-TB Magnetic Disk "Read" Heads *22*
2.4 Accelerometers and Semiconductor Lasers *24*
2.5 Nanophysics-Based Technology: Medical Imaging, Atomic Clock, Sensors, Quantum Computers *26*
References *27*

3 Systematics of Scaling Things Down: $L = 1\,\text{m} \rightarrow 1\,\text{nm}$ 29
3.1   One-Dimensional and Three-Dimensional Scaling   29
3.2   Examples of Scaling: Clocks, Tuning Forks, Quartz Watches, Carbon Nanotubes   31
3.3   Scaling Relations Illustrated by Simple Circuit Elements   37
3.4   Viscous Forces for Small Particles in Fluid Media   38
3.5   What about Scaling Airplanes and Birds to Small Sizes?   39
References   40

4 Biology as Successful Nanotechnology   41
4.1   Molecular Motors in Large Animals: Linear Motors and Rotary Motors   41
4.2   Information Technology in Biology Based on DNA   46
4.3   Sensors, Rods, Cones, and Nanoscale Magnets   52
4.4   Ion Channels: Nanotransistors of Biology   53
References   53

5 The End of Scaling: The Lumpiness of All Matter in the Universe   55
5.1   Lumpiness of Macroscopic Matter below the 10-µm Scale   55
5.2   Hydrogen Atom of Bohr: A New Size Scale, Planck's Constant   57
5.3   Waves of Water, Light, Electron, and Their Diffractions   60
5.4   DeBroglie Matter Wavelength   62
5.5   Schrodinger's Equation   63
5.6   The End of Scaling, the Substructure of the Universe   63
5.7   What Technologies Are Directly Based on These Fundamental Particles and Spin?   64
Reference   65

6 Quantum Consequences for the Macroworld   67
6.1   Quantum Wells and Standing Waves   67
6.2   Probability Distributions and Uncertainty Principle   69
6.3   Double Well as Precursor of Molecule   71
6.4   The Spherical Atom   73
6.5   Where Did the Nuclei Come From (Atoms Quickly Form around Them)?   75
6.6   The "Strong Force" Binds Nuclei   75
6.7   Chemical Elements: Based on Nuclear Stability   76
6.8   Molecules and Crystals: Metals as Boxes of Free Electrons   77
References   79

7 Some Natural and Industrial Self-Assembled Nanostructures *81*

7.1 Periodic Structures: A Simple Model for Electron Bands and Gaps *81*
7.2 Engineering Electrical Conduction in Tetrahedrally Bonded Semiconductors *83*
7.3 Quantum Dots *85*
7.4 Carbon Nanotubes *86*
7.5 C_{60} Buckyball *91*
References *92*

8 Injection Lasers and Billion-Transistor Chips *93*

8.1 Semiconductor P-N Junction Lasers in the Internet *93*
8.2 P-N Junction and Emission of Light at 1.24 μm *98*
8.3 Field Effect Transistor *101*

9 The Scanning Tunneling Microscope and Scanning Tunneling Microscope Revolution *105*

9.1 Scanning Tunneling Microscope (STM) as Prototype *105*
9.2 Atomic Force Microscope (AFM) and Magnetic Force Microscope (MFM) *110*
9.3 SNOM: Scanning Near-Field Optical Microscope *115*

10 Magnetic Resonance Imaging (MRI): Nanophysics of Spin ½ *117*

10.1 Imaging the Protons in Water: Proton Spin ½, a Two-Level System *117*
10.2 Magnetic Moments in a Milligram of Water: Polarization and Detection *118*
10.3 Larmor Precession, Level Splitting at 1 T *119*
10.4 Magnetic Resonance and Rabi Frequency *120*
10.5 Schrodinger's Cat Realized in Proton Spins *121*
10.6 Superconductivity as a Detection Scheme for Magnetic Resonance Imaging *122*
10.7 Quantized Magnetic Flux in Closed Superconducting Loops *123*
10.8 SQUID Detector of Magnetic Field Strength *124*
A SQUID-Based MRI Has Been Demonstrated *125*

11 Nanophysics and Nanotechnology of High-Density Data Storage *127*

11.1 Approaches to Terabyte Memory: Mechanical and Magnetic *127*
11.2 The Nanoelectromechanical "Millipede" Cantilever Array and Its Fabrication *127*

11.3 The Magnetic Hard Disk *132*
Reference *137*

12 Single-Electron Transistors and Molecular Electronics *139*
12.1 What Could Possibly Replace the FET at the "End of Moore's Law"? *139*
12.2 The Single-Electron Transistor (SET) *139*
12.3 Single-Electron Transistor at Room Temperature Based on a Carbon Nanotube *142*
12.4 Random Access Storage Based on Crossbar Arrays of Carbon Nanotubes *143*
12.5 A Molecular Computer! *147*
References *149*

13 Quantum Computers and Superconducting Computers *151*
13.1 The Increasing Energy Costs of Silicon Computing *152*
13.2 Quantum Computing *152*
13.3 Charge Qubit *154*
13.4 Silicon-Based Quantum-Computer Qubits *155*
13.5 Adiabatic Quantum Computation *157*
Analog to Digital Conversion (ADC) Using RSFQ Logic *159*
13.6 Opportunity for Innovation in Large-Scale Computation *160*
References *161*

14 Looking into the Future *163*
14.1 Ideas, People, and Technologies *163*
14.2 Why the Molecular Assembler of Drexler: One Atom at a Time, Will Not Work *166*
14.3 Man-Made Life: The Bacterium Invented by Craig Venter and Hamilton Smith *169*
14.4 Future Energy Sources *171*
14.5 Exponential Growth in Human Communication *173*
14.6 Role of Nanotechnology *175*
References *175*

Notes *177*
Index *199*

Preface

A revolution has occurred over the past several decades in the availability and uses of information. This is perhaps the strongest reminder that we live in a time of accelerating technological change. This book explains one aspect of technological change, related to very small devices, devices approaching the atomic scale in their size. The technology related to small devices is called nanotechnology. But our aim in this book is broader, to put nanotechnology into the context of earlier scientific advances concerning very small objects. The contributions of the enlarged field of "nanotechnology" have been particularly great in information technology, the technology of computers, wireless communication, fiber optics, the Facebook phenomenon, and thinking machines like the "Watson" computer that can win on the television game "Jeopardy." We will argue that the greatest success of nanotechnology is really the silicon chip, with its billion transistors. Although Moore's law appeared before the word "nanotechnology," these developments in silicon technology clearly now fall within the definition of "engineered systems, at least one dimension being in the scale from 100 to 1 nanometers." The best way to view these developments is as part of nanotechnology. Many people, we think, will benefit, beyond seeing that silicon technology is a leading example of nanotechnology, by recognizing the longer common thread of competences that we believe are best regarded as "nanotechnology." These have the common aspect of harnessing tiny objects, to include the electron spins in cesium atoms that are the basis for the atomic clock, the use of proton spins in successful magnetic resonance imaging, and other topics as we will mention.

Specifically, this small book was stimulated by the invitation of "The Modern Scholar" series of audio lectures of Recorded Books, LLC, to one of us to provide a series of audio lectures on the topic "Understanding Nanotechnology: A Bridge to the Future." We have benefited from interactions with many people in this project. We thank Ed White of Recorded Books;

Ed Immergut, Consulting Editor in Brooklyn, NY; Vera Palmer, Commissioning Editor at Wiley VCH; Ulrike Werner of Wiley-VCH; Prof. Lorcan Folan; and Ms. DeShane Lyew at the Physics Department of NYU-Poly. In particular, E.W. thanks Carol, Ph.D. in Mathematics and Prof. of Computer Science, for help in many ways and for comments on the abacus and more generally on the history of mathematical inventions. M.M. wants to thank her family and friends for their tremendous support.

The book is dedicated to pioneers in the nanotechnology-enabled information revolution. John V. Atanasoff invented the digital programmable computer, arguably the most important invention of the twentieth century, as detailed in Note N4 to Chapter 2. (*Notes* follow Chapter 14 in the organization of this book.) John Bardeen was a coinventor of the transistor, which made the digital computer a practical matter and led to the Moore's law growth of computing capacity. S.S.P. Parkin did essential developmental research allowing the quantum-mechanical magnetic tunnel junction, based on the spin ½ of electrons, to be manufactured as the data-reading element in today's computer memory, the basis for cloud computing. Sir Timothy John "Tim" Berners-Lee is a principal architect of the World Wide Web, the global computer network that connects people in today's world.

Brooklyn, NY, May 1, 2011 *E. Wolf and M. Medikonda*

1
Discovery, Invention, and Science in Human Progress

Nanotechnology is a recent addition to the long history of human efforts to survive and make life better. Nanotechnology is based on the understanding of and tools to deal with very tiny objects, down to the size of atoms [1]. To begin, it is worth reviewing some of the broader history, to put nanotechnology in perspective, so that we can better understand how it can serve as a bridge to the future.

Technology has evolved over tens of thousands of years and more by the activities of humans and their predecessors: the history of technology is almost the history of humanity.

1.1
Origins of Technology, the Need for Human Survival

Struggling for survival and ascendency for over 50 000 years (a conventional time frame for the migration of "homo sapiens" out of Africa [2], (see Figure 1.1), humans invented new and useful ways of doing things.[1] Technology has advanced ever since, in an accelerating fashion, and we hope to provide an understanding of a current forefront of technological advance called *nanotechnology*, which specifically deals with small objects and the laws of nature that describe these small objects [1].

Technology, often based on discovery, is knowledge on how to get things done, and the tools to make use of that knowledge. This is a practical matter, often a matter of life and death. *Stone age* tools have been found dating to about 2.4 million years ago. Then came the *Bronze age* and the *Iron age*. In 1200 BC, the Hittites were the first to use iron in weapons. We can say that advanced metal technology started long ago [3–7].[2] To understand nanotechnology it is useful to review some of the previous technological advances in the 50 000-year history.

Understanding the Nanotechnology Revolution, First Edition. Edward L. Wolf, Manasa Medikonda.
© 2012 Wiley-VCH Verlag GmbH & Co. KGaA. Published 2012 by Wiley-VCH Verlag GmbH & Co. KGaA.

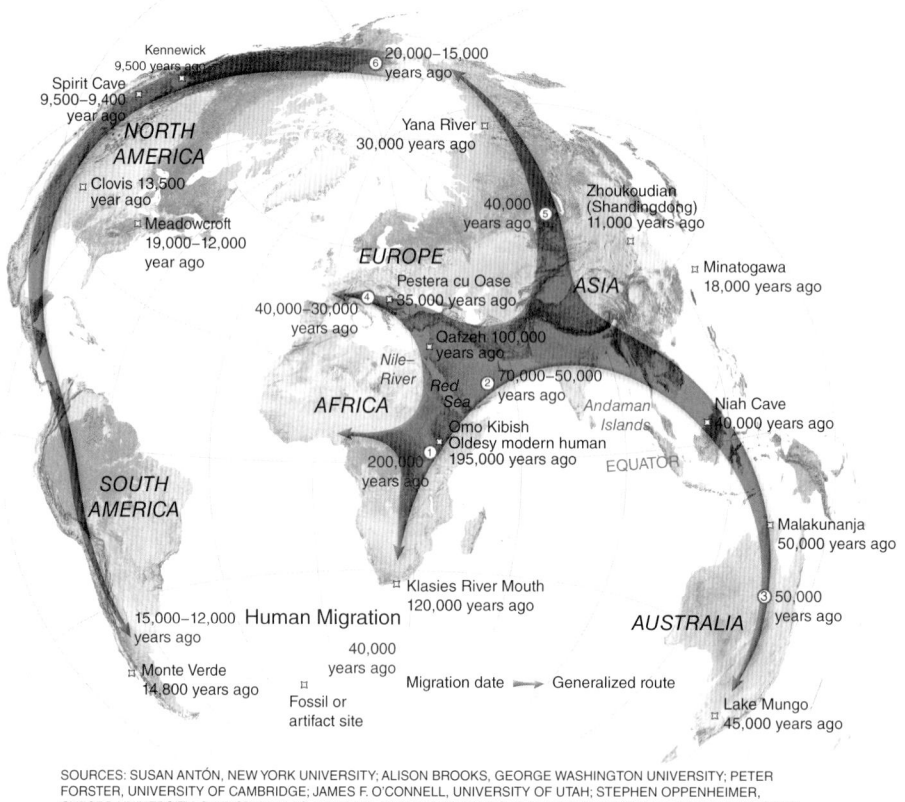

Figure 1.1 A speculative *but data-based* map of human migrations, from genomic technology. *Homo sapiens* migrations, with approximate dates in thousands of years, are tracked by changes in human DNA. We discuss this in Chapter 4. (NG Maps/National Geographic Stock).

1.2
The Industrial Revolution: Watt's Steam Engine, Thermodynamics, Energy Sources

The development of the wheel, advanced control of fire, and the development of copper, bronze and iron technologies, set the stage for the more recent industrial revolution. The industrial revolution, based on the invention of the steam engine by James Watt in 1776, led quickly to the steam locomotive in 1804. This required a *synthesis* of the technologies for making fire and elaboration of wheels and axles to include gears and pistons, requiring knowledge of metals to make strong components. The steam engine also brought to the fore knowledge of thermodynamics, a science that could improve the efficiency of engines based upon steam. The concept and measurement of temperature, an aspect of modern science, was part of that advance.

The advance of civilization can be measured by the technology in use and also by the sources of energy that were available at a given time.

A primary source of energy in the mercantile sailing era was wind. Wind has been used since ancient times, to make sailing boats and to power windmills to pump water or grind grain. It is reported that in fourth century BC the Greek wind- and human-powered merchant fleet went all the way from Spain to the Black Sea, and of course Julius Caesar invaded Egypt by sea. A three-masted merchant ship is reported in China in 400 AD. The *compass*, based on magnetite, an iron oxide, was invented in 200 BC in China, and was widely used by the Chinese shipping fleet in 1100 AD. Sailing long distances stimulated the development of better clocks, needed for navigation, and of course, clocks are important in today's information technology.

The technology of sailing ships and worldwide navigation flourished starting from the time of Columbus, who sailed in 1492 to America from Spain. In 1503 Vasco Da Gama of Portugal took 20 ships from Lisbon, around the bottom of Africa and to India, initiating more wide-ranging open-sea commerce, which had earlier been limited, such as to the Mediterranean Sea. Sailing ships remained important until well after 1860, when steam-powered ships were first constructed.

The Dutch were well known for pumping water with windmills, the predecessors of modern wind turbines. Present-day technologies building on the sailing era technology include airfoils on airliners and the space shuttle, helicopters, and wind turbines of 1 MW (megawatt) generating capacity, that cost about $1 M apiece.

With James Watt's invention of the steam engine in 1776, to begin the industrial era of engines, the source of energy shifted, from wind to fuels to be burned to generate steam and run the engine. Over time, the fuel of choice has changed from wood, to peat and coal, and then to oil and gas. A recent addition is nuclear power, used by nuclear reactors in submarines, aircraft carriers, and electric power plants. This might be looked at as passing industrial leadership from Holland (wind technology) to England (wood and coal steam engines) and then to America (the era of oil and gas, Henry Ford, and the internal combustion engine). Nuclear energy, since about 1945, with the first nuclear reactor in Chicago developed by Enrico Fermi, has been an international effort.

Although oil was well known in the Middle East since very early times, the modern large-scale extraction of oil as a fuel dates to 1859, in Titusville, Pennsylvania and 1901, in Spindletop, Texas. US oil production peaked in about 1971, and, with depletion, has fallen ever since. The era of availability of oil may be about 200 years, starting in 1859, because the amount of available oil is definitely limited.

1.3
A Short History of Time: Navigation, Longitudes, Clocks

A short history of time involves the technology of devices to measure intervals of time. The earliest clocks were water clocks that date back to the sixteenth century BC in Babylonia and Egypt. These simple useful devices are similar in principle to the sand hourglass, depending upon a steady flow rate of a given mass of water or sand. Accuracy and resolution in clocks was stimulated by the need to know the *longitude* when crossing an open sea, far from sight of land. The distance in going from one time zone to the next is 15 degrees longitude, which is 645 miles at the Latitude of London, England. This information could be used by the sea captain. Suppose, at the wharf in London, as he sets sail, the captain's ship clock reads noon when the sun is directly overhead.

After a day of sailing to the west, noon the next day, the sun might be directly overhead at 12:30 on his ship clock (which was set in London). If so, the captain would know, assuming constant latitude, that he had traveled half a time zone, about 322 miles.

For lack of good clocks, this option was not available to sea captains until after 1760, with the invention of an accurate portable clock, the "marine chronometer," by John Harrison. For centuries before this, sea captains, in practice, had relied on *dead reckoning*.[3] The ship's compass indicated the direction of travel, and the distance per day was estimated from the speed multiplied by the time elapsed. This was a laborious and honest, but inaccurate process. Ships went astray and lives and fortunes were lost. The failings in navigation became such a problem that the British government established a Board of Longitude, to fund the development of an accurate clock.

A great advance came in 1759 with the invention of the accurate marine chronometer. This clock, based on a spring oscillator, was invented in stages by John Harrison [8], who won a prize from the British government. Somewhat earlier, in 1656, the pendulum clock was invented by Huygens. The idea of a pendulum as establishing a timescale was known earlier, even to Galileo in the early 1600s. A clock based on a pendulum was not built, however, until 1656. Although it predated the John Harrison chronometer, the pendulum clock is not useful except in fixed locations. It requires a stable footing not available on a ship.

A major advance in modern timekeeping was made with the miniaturized quartz oscillator in the Bulova watch (see Chapter 3). Here quartz is shaped to form a cantilever or spring, whose resonant frequency, f, near 33 kHz, is governed by a formula $f = (1/2\pi)(K/m)^{1/2}$, where the spring constant K has units Newtons/meter and m is the moving mass. (Here, the Newton is about 0.225 pound force, and m is measured in kilograms. The frequency is in hertz, oscillations per second.) The quartz oscillator has been further miniaturized and still forms the clock in the personal computer (PC), working up to 3 GHz, as we will discuss in Chapter 3.

The *atomic clock*, based Cs (cesium) atoms, is now used as the worldwide standard of timekeeping, accurate on a scale of nanoseconds, billionths of a second.

Our book is about *nanotechnology*, the useful (and profitable) application of small-scale working elements and devices [1].

Present major technologies that benefit most from nanotechnology are the *silicon computer technology*, and *information technology*. Information technology (IT), couples silicon technology with optical fiber transmission of signals, and with advances in data-storage technology, such as the computer disk drive.

Medical technologies including diagnostics such as X-rays, which are also the basis for unraveling the structure of double-helix DNA (the information aspect of biology), drug design and genomic technology; and magnetic resonance imaging (MRI) also benefit from the emerging area of nanotechnology.

High-energy synchrotron light sources, adapted from high-energy physics, giving huge intensities of X-rays, have allowed rapid determinations of molecular structures. This has enabled modern pharmaceutical advances, which also benefit from computer modeling. While polymers (think polyethylene and polystyrene) are definitely chemistry, one can argue that designing drugs for a specific purpose is an exercise in nanotechnology.

1.4
The Information Revolution: Abacus to Computer Chips and Fiber Optics

The advance of technology is itself accelerating [9]. To illustrate this, we will consider the timing of advances related to information technology! The first record of bones carved with notches dates to 20 000 BC, and bones carved with prime numbers were found as early as 8500 BC. (Prime numbers cannot be expressed as a product $c = ab$ of two other numbers. This is a subtle matter, but evidently understood by smart people nearly 10 centuries ago.) The abacus comes from China and Babylonia, around 1000 BC, and has several forms.

But it was not until 1500 AD that Leonardo da Vinci described a mechanical calculator. Logarithms and the slide rule were invented about 1600 AD. The predecessor of the IBM tabulating machine was invented by Hollerith in 1890 for the US census. (If time starts 50 000 years ago, then the 411-year interval (2011 AD–1600 AD) is 0.8% of the life of Homo sapiens. If time starts 4.54 billion years ago, with the formation of the earth, this is in the last 0.09 millionth of time on Earth.

Time intervals between inventions in this set have reduced from thousands of years to hundreds of years.

But since 1945 or so, a period of 65 years, we have had many, many inventions related to information technology! The transistor was invented in 1947, the Univac programmable computer in 1951, the atomic clock in 1955, the integrated circuit in 1958, the Xerox machine in 1959, the laser in 1964, the magnetic floppy disk in 1971, the Ethernet in 1973, the personal computer in 1973–1976, the optical fiber in 1970–1975, the injection laser in 1978, the Internet global computer network in 1990 and the Pentium chip in 1993, global positioning system (GPS) in 1993, the Internet search engines in 1993–1998, the Blue Gene chess-playing computer in 1997, the magnetic tunnel junction hard disk reader in 2004, and Watson computer winning the Jeopardy TV competition in 2011.

While earlier inventions were spaced by hundreds or even thousands of years, the inventions listed here since 1945 are spaced by about 4 years, a much shorter interval! It is widely agreed that technology is accelerating [9]. Moore's law, which has predicted the doubling of the number of transistors per chip each 1.5 years, is an example of "exponential growth," an accelerating increase. A striking, but hypothetical scenario on growth of computing capacity is "the Singularity," when computer intelligence may exceed human intelligence. It is suggested, in the work of Ray Kurzweil [9], but hotly debated that computers will become completely equal to humans in all thinking activities in 2045. "Singularity" or not, we live in an age of accelerating technological capacity, much of it based on nanotechnology, the topic of our book.

1.5
Overlap and Accelerating Cascade of Technologies: GPS, Nuclear Submarines

In a more definite historical sense, we have seen that combining two technologies can lead to a third, for example, combining use of fire and wheels led to the steam engine. (Actually, more developments were needed, including the mining and purification of metals, and their fabrication into gears, pistons, and more.) We can think of the steam engine as a *hybrid* technology, and really this is the norm in the advance of technology. What we see is really a cascade of developments, one upon the other. A recent example is the global positioning system (GPS) depending on atomic clocks, state of the art silicon devices, computers, and space technology.

Another hybrid technology, with a strong nanophysical component, is represented by the nuclear submarine. (We use the term *nanophysics* for the forms of physics that are needed to describe properties and processes on *size scales below 100 nm*; these include atomic and solid-state physics, and nuclear and high-energy physics.)

The nuclear submarine uses an onboard nuclear power reactor to allow long voyages, and nuclear-powered submarines and aircraft carriers have been extremely reliable. The power that drives the propellers on the submarine comes from an electric motor. The electric motor is run from an electric generator. The generator is turned by a steam turbine. The heat that generates the steam comes from a nuclear reactor. The nuclear reactor derives its energy from fission events within uranium nuclei (see Chapter 5). The prototype reaction is the splitting of the ^{235}U, which has $Z = 92$ protons and $N = 143$ neutrons, to release energy of several million electronvolts (MeV) per nucleus. (This is a huge energy release; it is about a million times larger than the energy release in a chemical reaction such as burning hydrogen to make water.) The energy release comes because the Coulomb repulsion between positive proton charges in the resulting nuclei, which might be typified by ^{133}Cs (cesium), with proton number $Z = 55$, is lower after the fission.[4] The difference in energy can also be calculated by using Einstein's famous relation that the change in energy will be the change in mass multiplied by c^2, with c the speed of light at 2.998×10^8 m/s.

So the nuclear-powered submarine is an example of cascading of technologies, and the origin of its power emphasizes the importance of nanophysics in our modern world.

Understanding the nanophysical properties of the nucleus slowly accumulated by the work of many physicists over decades.

1.6
Silicon and Biotechnologies: Carbon Dating, Artificial Intelligence

Other useful technologies that stem from the knowledge of nanophysics of atomic nuclei include the technique of "carbon dating," which allows us for example, to date artifacts such as bones of dinosaurs and wooden tools and remains left by early humans. Carbon dating depends upon the presence of small amounts of ^{14}C in the air. The stable form of carbon is ^{12}C. The carbon isotope ^{14}C decays on a regular timescale, and a living organism has its ^{14}C replenished by respiration. When it dies the respiration stops and the ^{14}C decays. The amount of ^{14}C remaining can be determined by counting the rate of electron emissions per second per gram of carbon in the specimen. This knowledge, on which the technology of carbon dating rests, has accumulated slowly by the activity of nuclear physicists. This technology was used, for example, to establish the date of manufacture of the bone flute recently found in a cave in Eastern Europe. The flute was found to have been made 30 000 years ago, from the hollow bones of a large bird. The flute had carefully chiseled openings to place the fingers, and a headpiece to create sound from blowing air.

We suspect that one source of the accelerating advance of technology is the greatly enhanced exchange of information [9]. The Internet global computer network allows the exchange of ideas in an unprecedented fashion. This builds on the silicon technology, the technology for storing information, the optical fiber technology for communications all around the world, and the advance in computer programming.

Human capability and human productivity is basically multiplied by technology. It is definitely rising with the increase in communication capacity. With the Internet, the global computer network, an idea from a remote researcher is often put onto the Web, the Internet, as an Internet archive file that immediately can be read by people all around the world. This is the democratization of science, unquestionably allowing faster progress. There are smart, well-informed people all over the world, and now more of these people can participate in the creation of new knowledge and new technology.

Consider the Google Book project, the ongoing effort of Google Corp at digitizing all the books in the world, to make the texts available as searchable computer PDF files for everyone. There are millions of books in libraries. However, the old books, especially, have been inaccessible, on the whole, limited to those who can walk, for example, to the Library of Congress or the Stanford University Library. Interlibrary loan of books is possible in principle, but difficult in practice, especially for a rare old book.

The Internet ("World Wide Web") and the Google Books project are revolutionizing access to books. Google Books have digitized millions of books, which are available on the Web, with the great advantage that the digitized sources can be *searched* in a flash (ask the Jeopardy contestants). In his extended family, one of us, EW, had shared a single copy of the "Burgess

Genealogy (1865)," which he finally photocopied. But now photocopying is becoming obsolete, because the "Burgess Genealogy" is on Google Books and can be freely downloaded as a PDF file by anyone. The great advantage is that the computer file can be searched, to find where a given word appears in the book. For example, a search for "Thomaston" immediately finds which relatives named Burgess lived in Thomaston, Maine (William Carey Burgess). There is no need, now, to travel to the one or two libraries in the country that have this book; it is easily available to anyone in a minute on his computer.

EW wrote a book "Quantum Nanoelectronics" [10] for use in his university teaching. This is a long book with facts sometimes forgotten, for which it is a good source. It turns out that the quickest way to find a fact or number from this book is to open the book on Google Books and search it. Putting in "Bohr magneton" (you can try it) gets instantly eight responses, and one can scroll down and see each of the pages that contain that phrase. It is much faster than looking through the index! The use of Google Books online search clearly is easier, once the computer is online, than fumbling through the index and then flipping pages.

This really useful facility is now available for millions of books, a substantial fraction of books in the major libraries in the United States. This provides a big change in the working conditions for scholars worldwide (and for ordinary people who may just want to find their ancestors).

This is a part of the ongoing information technology explosion, which has strong roots in nanotechnology. It seems sensible to call the Google Book search capability a form of *practical artificial intelligence*, or AI from a practical point of view. It does a task that a human might otherwise be needed to perform.

In another example of such *practical artificial intelligence*, Raymond Kurzweil invented a reader for the blind. Given a sheet of text the reader will absorb the information and speak it out in English or any desired language! A complementary capacity, *speech recognition* (for which software is available, for example, from the Nuance Corporation), will take spoken words and turn them into a Word file.

A doctor may use such a program to get a written record of his discussions with patients. The accuracy of such programs is definitely improving. So a computer can now take your spoken words and turn them into a file and a written page. A computer can also look at a written page and speak it out in any language you might like. These activities are valuable and would in earlier times require a human to carry them out. We consider these to be examples of *practical artificial intelligence*, and my inclination would be to call this simply AI, for artificial intelligence.

Other people disagree, and offer a different definition of (strong) artificial intelligence, AI, which does not exist as far as we know.

This hypothetical form of AI, which we will call *strong* or *sentient* AI, would be a self-aware conscious independent entity that some people think (and/or fear) might arise from a computer system. This hypothetical kind of artificial intelligence, AI, was described long ago by Isaac Asimov in his book entitled "I, Robot." Asimov's hypothetical robot knows who it is and will defend its own interests!

If a big robot looking like Arnold Schwarzenegger strides into the room and says "I am in charge here," you will know that *strong AI* has arrived.

Asimov's robot has not happened yet. This "strong" or "sentient" AI has not occurred, but it is part of the Singularity hypothesis. (The machine called "Watson" that has recently won on the TV show "Jeopardy" is not likely to stride into anyone's room, since it comprises two rooms of computer racks, costing millions of dollars [9] and requiring megawatts of power to run.)

But the practical form of artificial intelligence is certainly helping the acceleration of technology. This is part of what we are talking about. All of the elements of the practical form of artificial intelligence incorporate nanotechnology: Fast computers with lots of storage, fiber-optic networks connecting to cloud computers all incorporate basic nanotechnological inventions.

One of us was driving a rental car in summer 2010 in Colorado, using a GPS unit carried in a carry-on bag from New York. On the highway, a casual query to the GPS about "driving home" (to a New York apartment) led quickly to a voice from the GPS describing the requested trip as "1875 miles and would take 38 hours." This is pretty amazing; it did a search of all the routes from an immediate location in Colorado, using maps stored in its extensive database. The selected route was measured and an estimate of the time was produced, all in a few seconds. (The spoken answer could as easily have been given in German or in Hindi, according to the manual with the inexpensive GPS device.)

This seems to be an extraordinary artificial intelligence! Nanotechnology enters this story in fast computing, large databases, and in the atomic clocks in the GPS.

The New York Times recently ran a story [11] about a commercial computer program, called "Creative Artificial Intelligence (CAI)," for writing advertisements or ads. This program, developed by an arm of Havas, an international communications group, will produce ads on request, based on the selection of a product category (e.g., "soft drinks") followed by a brief dialog. This CAI software can generate an estimated 200 000 ads.

In a recent demonstration, according to the New York Times story, the software "brought forth bland and formulaic – but perfectly acceptable – ads that could be run in magazines or newspapers, or as banners on Websites or billboards." The story [5] was not particularly complimentary, describing as "mediocre" ads produced in a typical ad agency and also as "mediocre" ads produced by the software program that was being described.

With regard to reactions from people who did trials in use of the software, trial users were, generally, at first, amused. But, the notable quote was "they get a little scared when they see that a software program can create the same (mediocre) results *in just 10 seconds* as several hours of strategic meetings and production" in an agency! *(italics added)*.

It was commented that this CAI program "generates fun ads," and also that it "sometimes leads to random accidents that could stimulate the creative process."

Nanotechnology enters this scenario in fast computation and rapid access/search to a large database; the large database was probably collected by using Internet search as pioneered by Google, which again has roots in nanotechnology.

A program to generate classical music has been written by Prof. David Cope, a music professor at the University of California. As in the advertising software, the style of the requested music can be adjusted (Bach or not Bach?). The current version of the software is called "Emily Howell," an earlier version was called Emmy, and the output of new music written by these programs includes *1500 symphonies, 1000 piano concertos and 1000 string quartets*! An article on this subject (*I'll be Bach*) [12] written by Chris Wilson appeared in Slate, dated May 19, 2010.

The impact of nanotechnology in the present and future, we argue, comes in part through its participation in the technology of the information explosion,[5] and a leading example is artificial intelligence of a practical kind. A further example of nanotechnology-enabled practical artificial intelligence, in the form of the IBM Deep Blue computer, reached a peak when it outplayed the chess master Gary Kasparov. More recently, IBM has coupled its most powerful computer, Blue Gene, to a huge database, equipped with algorithms to direct extremely fast searches as on Google, to make "Watson," the supercomputer that can play the TV game Jeopardy in a winning fashion [3, 7].

When human contestants in test trials play Jeopardy against "Watson," they tend to refer to "Watson" as He. You may ask, "Who is 'Watson?'", and, of course, "Watson" in reality is a roomful of advanced computers [13], very expensive, with a large electric power bill, and there is absolutely no chance

Figure 1.2 The show Jeopardy in February 2011 pitted Watson against two star players, Ken and Brad. Watson won by the numbers, but made a few spectacular goofs, including placing Toronto, Canada, in the United States [14].

of "Watson" reproducing. (Although a "Watson"-type machine might be able to *design* a better version of itself) (Figure 1.2).

In modern business practice, this type of capacity is called an "expert system," designed to serve as a consultant, for example, to inform decisions in business or medical diagnosis. IBM's "Watson" programmed supercomputer/database is at the apex of the development of such expert systems [9]. Clearly the advance of Moore's law will make this capacity more available in the future. At the moment "Watson," however, is a multimillion dollar capital item, although time sharing may well evolve as a practical alternative.

Clearly there is a big disconnect between a hypothetical robot looking like Arnold Schwarzenegger and the real IBM "Watson," which is a roomful of computers completely tied to the power grid. So even if "Watson" should decide that "he is in charge," "he" would be immobile, and his owners at IBM could unplug him if they felt he was getting out of hand.

By any measure [1], the Pentium computer chip and its billion-transistor successors, used by the millions (nearly a thousand servers) in a system such as the IBM Watson, are products of nanotechnology. Open access to, and the facility of rapid searching, much of the knowledge of our civilization, is a recent and ongoing development. It is hard to escape the conclusion that will have an astounding effect in technological advance.

Also, in many countries, detailed reports were made freely available to public listing out the risks and benefits and the areas of technology that still need investigation. And a lot of investment is being put into nanotech toxicology research to fully understand the toxicology implications of the materials [15].

1.7
Nanotechnology: A Leading Edge of Technological Advance, a Bridge to the Future

What is nanotechnology? It is the knowledge and methods to deal with and engineer particles and systems that range from 100 nm in size down to the sizes of individual particles, including, from our point of view, the particles in the atomic nucleus. The definition offered by the National Science Foundation in the United States describes controlled engineered elements whose extent at least in one dimension is in the range of 1–100 nm. The Royal Society Report [1] extends the range to 0.2 nm "because at this scale the properties of materials can be very different from those at larger scale." This report [1] (p. vii) also comments that "nanotechnologies have been used to create tiny features on computer chips for the past 20 years." This report clearly identifies the information revolution as a beneficiary of nanotechnology. We are reminded of the reality of this revolution by the recent valuation estimates of Facebook, *ca* $50 billion, and by a report in the Egyptian newspaper "Al-Ahram" that an Egyptian father, Jamal Ibrahim, had named his newborn daughter "Facebook" to honor the social media site's role in Egypt's revolution [14].

Conventional engineering, based on machine tools making three-dimensional objects, such as tiny Ti screws for dental implants, really extends only to about 1 mm, the size of parts in mechanical wristwatches or sophistical dental implants made of titanium.

There are six orders of magnitude, a factor of 1 million, in length from 1 mm to 1 nm. These size scales, from 1 mm to 1 nm, largely remain to be exploited to make useful devices and systems. In this context, the physicist Richard Feynman said, "there is lots of room at the bottom." The potential benefits of miniaturization are illustrated by Moore's law, which describes the road to billion-transistor chips at continuously falling prices and vastly improved chip performance. Consider that semiconductor counters now work at frequencies up to 200 GHz! (3 GHz is 3 billion cycles per second and is now a typical clock frequency in personal computers.)

But most areas of technology remain to be extended into the nanotechnology domain.

We believe it is reasonable, certainly for the benefit of the readers of this informational book, to *broaden our view of nanotechnology* to include devices and systems whose operation is based on the quantum laws of nature, which we call nanophysics, which come into play on size scales below 100 nm and apply all the way down to 1 femtometer, 10^{-15} m, one millionth of a nanometer.

Examples of nanophysically based devices include the magnetic tunnel junction disk reader, leading to 1-TB hard disks (1 TB is 10^{12} bytes), the cesium atomic clock used in the GPS and in Wall Street computers, and the MRI (magnetic resonance imaging) medical imaging apparatus, whose operation is based on the nanophysics of a spin-½ particle, the proton.

The laser, powering optical fiber communication, is based on the quantum photon of light and on the quantum phenomenon of stimulated emission first described by physicist Albert Einstein in 1905. These developments are based on quantum physics, the underlying science of nanotechnology. Nanotechnology, in its strongest form, would allow engineering of devices in this entire size range. It would be engineering knowledge applicable to size scales from 100 nm down to 1 femtometer, the size of the nucleus of an atom. Although femtometer engineering is not possible, intelligent applications of femtometer systems as in magnetic resonance imaging (MRI) and nuclear reactors can be viewed as "nanotechnology."

In our beginning chapter, we have sketched a review of the history of technology, learning that technology is an accumulating bank of know-how and devices that various aspects of technology combine or form hybrids. We have seen that technological advance is accelerating. Nanotechnology is expertise and methods relating to small-size objects, on sizes below 100 nm, or about 0.1% the size of a human hair. A leading example of nanotechnology is the silicon chip that may contain a billion units that operate at frequencies above 1 GHz, a billion operations per second. It is reasonable to include in our discussion of nanotechnology those devices that depend upon engineering knowledge of atomic-scale phenomena, including electron and proton spins, in modern disk-drive readers and in magnetic resonance imaging (MRI), respectively. Nanotechnology has been, and will continue to be, a leading part in the advances in communication and computing, components of the information technology and Internet advances that are transforming our way of life.

Our hope in writing this book is to promote broader understanding of the existing nanotechnology and its underlying nanophysics, and its role in the ongoing information revolution, and to help show how it may provide a bridge to the future.

1.8
How to Use This Book

The central theme of our book, stated in Chapter 1, continues in Chapters 2, 3, 5–7, and 14. Important illustrations of the central theme are given in Chapter 4 and Chapters 8–13. Chapter 4, on biology as successful nanotechnology, and Chapters 8–13 outline the present state of nanotechnology in several important areas. These chapters are factual and reasonably independent of one another. For the reader who wishes to sample highlights among these, we suggest that Chapter 4 stands apart.

Scaling to small sizes is central to our topic, and involves some mathematics. Chapter 3 shows on a case-by-case basis how the operating frequency f of oscillators (and other devices) increases as the typical size scale L is reduced. If the numerical examples in Chapter 3 become tedious, be sure to look at Figure 3.5, which shows how typical frequencies increase from 0.5 Hz for a grandfather clock to 10^{14} for a hydrogen molecule as a mechanical oscillator. (Don't miss Figure 3.1 either.) The latter portion of Chapter 3, explaining why small objects like pollen particles float and don't follow trajectories like baseballs, may be more mathematical and optional, unless you are curious as to why a bumblebee can't "fly" in an aeronautical sense. Chapters 8, 9, and 11–13, illustrative of our central theme, describe the most important present and future devices of electrical nanotechnology. These topics are up-to-date and are introduced with the minimum mathematics to convey their essence. Chapter 10, however, is a little different, it explains how the MRI machine works and how it is based on the quantum mechanics or "nanophysics" of the proton spin 1/2. Chapter 10 also explains how a less costly MRI machine can be made using a quantum-mechanical sensor device, the "superconducting quantum interference detector" (SQUID). (We include quantum mechanically based devices in our broadened category of nanotechnology.) The operating element of the MRI machine is the proton, whose radius is about a femtometer, one millionth of a nanometer. Chapters 7 (including carbon nanotubes and C_{60} molecules), 12, and 13 cover in some detail topics highlighted in the new book by Michio Kaku, "Physics of the future: How science will shape human destiny and our daily lives by the year 2100" (2011), which is part of our concluding discussion in Chapter 14. Our comments on the hypothetical "Singularity" (which may already have happened?) are limited to Chapter 1 and Chapter 14.

References

1. The Royal Academy of Engineering, The Royal Society of London (2004) Nanoscience and nanotechnologies: opportunities and uncertainties. Royal Society Report 2004, http://www.nanotec.org.uk/finalReport.htm (accessed 3 December 2011).
2. The long march of everyman, The Economist, December 20, 2005; The soul of an old machine: genomics is raising a mirror to humanity, producing some surprising reflections, The Economist, June 17, 2010.
3. Healy, M. (1993) *Qadesh 1300 BC: Clash of the Warrior Kings*, Osprey Publishing, Oxford.
4. Manning, J. (2011) Beyond the pharaohs. The Wall Street Journal, March 19, 2011.
5. Atkinson, Q. (2011) *Science*, **332**, 346.
6. Iskender, I. The Hittite Empire, http://iskenderi.tripod.com/id25.html (accessed 4 December 2011).
7. Wadley, L., *et al.* (2009) *Proc. Natl. Acad. Sci. USA*, **106**, 9590.
8. Sobel, D. (1995) *Longitude*, Walker and Co., New York.
9. Havelock, R.G. (2011) *Acceleration: The Forces Driving Human Progress*, Prometheus Books, Amherst, NY. Kurzweil, R. Wall Street Journal, February 17, 2011; *The Singularity Is Near: When Computers Transcend Biology*, Viking Penguin, New York, 2005.
10. Wolf, E.L. (2009) *Quantum Nanoelectronics*, Wiley-VCH Verlag GmbH, Weinheim.
11. Elliott, S. Don't tell the creative department, but software can produce ads too. New York Times, August 27, 2010.
12. Wilson, C. (2010) I'll be Bach. Slate Magazine, May 19, 2010.
13. Thompson, C. (2010) What is I.B.M.'s Watson? New York Times, June 16, 2010.
14. Watson wins "Jeopardy!" The IBM challenge. Culver City, CA, February 16, 2011, http://www.jeopardy.com/news/IBMchallenge.php (accessed 3 December 2011).
15. Bijker, W. None said "there are risks, so let's stop it." The Hindu, January 13, 2011, http://www.thehindu.com/todays-paper/tp-features/tp-sci-tech-and-agri/article1088455.ece (accessed 4 December 2011).

2
Smaller Is More, Usually Better, and Sometimes Entirely New!

As we learned in Chapter 1, nanotechnology is a key part of many recent advances, particularly in the technology of information, which are revolutionizing human existence.

Nanotechnology is based on making devices smaller. The most obvious advantage in making essential devices smaller is that more devices can be included, to multiply performance.

A second advantage, which is not obvious, is that usually the miniaturized device will work more rapidly and will cost less per unit.

It is now said that transistor production is "exa-units per year at nanodollar cost" [1]. This can be translated: "exa" means 10^{18}, which is one billion times one billion: One billion billions of transistors are manufactured per year! A "nano-$" is $/billion, or one billionth of a dollar or one cent divided in 10 million parts. This means 10 million transistors in production cost one penny. These nanotechnological devices have transformed our information technology and more.

Finally, the very process of making devices smaller has, in several cases, uncovered entirely new approaches based on quantum physics, or nanophysics, and new devices have been invented and developed.

The best example of a new-concept device may be the reader or sensor in computer magnetic disk drives, which now are available to store 1000 GB (1 GB = 1 × 10^9 bytes) at a cost of about $100. For comparison, the information totally contained in the Library of Congress has been estimated as 10 TB (1 TB = 10^{12} bytes)! We will learn more about this device in Chapter 11.

This is such an astounding achievement that it may help to put in a context. Think of a planar array of 1 cm × 1 cm squares, similar in scale to the beads on an abacus. So we imagine representing 1 bit by a 1 cm cube that is occupied or not in a binary sense.[1] How large an "abacus" would be needed to store 10 TB, a conventional measure of the information in the Library of Congress? (A *binary abacus* could in principle consist of N beads, each able to assume two positions, up or down, on a supporting wire. The amount of data would be 2^N.) To store 10 TB, one would need an (1 cm thick) "abacus" of side $L = 55.6$ miles.[2,3] The same content can now be stored in 10 laptop computers.[4]

2.1
Nanometers, Micrometers, Millimeters – Visualizing a Nanometer

Nanotechnology by name involves the nanometer, 10^{-9} meter. It is hard to visualize one billionth of a meter. We need to get some feeling about the size scales, what we are talking about when we say "small." If we imagine a meter stick, 39.37 inches: to get to a nanometer, can be thought of as first going to a millimeter (dividing by 1000), again dividing by 1000 to get to a micrometer, and then finally dividing by 1000 to get to 1 nm.

The resulting nanometer, 10^{-9} m, is about 10 times the size of the smallest atoms, such as hydrogen and carbon. But, again, the size of an atom is not familiar, and an atom maybe is better understood starting from the bottom, thinking of uniting electrons and protons.

A millimeter, the size of a pinhead, is roughly the smallest size available in present-day machines. The micrometer, at 0.1% of a millimeter is easier to visualize than a nanometer. A millimeter would be divided into a thousand parts to get to a micrometer. A micrometer is barely larger than the wavelength of visible light, too small to see.

A fine human hair is about 60 μm in diameter, so 0.06 mm is easy to see. From the size of a human hair to get down to a nanometer, we need to divide by 60 000. To get from 60 μm to 1 nm another way would be to cut it in half, once, twice, to 16 times. Taking a fine human hair and halving it 16 times will lead to 1 nm.

Similarly, if we think of a meter stick and cut it in half, then cut one of those pieces in half, in succeeding bisections, 30 are required to reach 1 nm. This is equivalent to saying that 2 to the power 30 is a billion.

One benchmark is the wavelength of visible light, about 500 nm or 0.5 μm. As we know by dispersing light in a prism or in a rainbow, white light contains varying wavelengths, a central value is near 500 nm. Another small object is a pollen particle, a tiny seed; it has to be small enough to float in air. Pollen particles that make our eyes water are on the size scale of 50 μm.

A final benchmark at small size is the atom, about 0.053 nm in radius. So, about 10 hydrogen atoms in a line would extend to 1 nm. Of course one cannot see atoms. Modern instruments including the transmission electron microscope (TEM) and the scanning tunneling microscope (STM) can image atoms, which makes one believe that they indeed exist, but does not help to intuitively understand a nanometer size. But it is possible to learn about atoms and then to think of them as an anchor or benchmark on the small-size end.

2.2 Moore's Law: from 30 Transistors to a Billion Transistors on One Chip and Cloud Computing

Miniaturization has been at the heart of information explosion. We believe that the Moore's law miniaturization is the first and best example of the tremendous advance making devices smaller. Moore's law describes the approximate doubling in the number of active units in the semiconductor chip every 1.5 years or so. This has been a continuing exponential growth.

The computer chip is a preeminent accomplishment of twentieth century technology, making vastly expanded computational speed available in smaller size and at lower cost. Computers and e-mail communication are almost universally available in modern society [2]. Perhaps the most revolutionary results of computer technology are the universal availability of e-mail to the informed and at least minimally endowed, and magnificent search engines such as Google. Without an unexpected return to the past, which might renounce and roll back this major human progress, it seems rationally that computers, fiber-optic communication, and advanced computer technology have ushered in a new era of information, connectedness, and enlightenment in human existence. An enthusiastic endorsement of this view has been offered by Havelock and many details by Kurzweil [3].

Moore's empirical law summarizes the "economy of scale" in getting a multiplied capability (device count) in the same size package by making the working elements ever smaller. (It turns out, as we will see, that "smaller" usually means faster, characteristically enhancing the advantage in miniaturization.)

In silicon microelectronic technology an easily produced memory cell size of $1\,\mu m$ corresponds to 10^8 bits/cm^2. Equally important is the continually reducing size of the magnetic disk memory element (and of the corresponding read/write sensor head) making possible the $\sim 100\,GB$ disk memories of contemporary laptop computers.

In 1965 a silicon chip contained perhaps 30 transistors. In the vicinity of 1990 the chip contained a million transistors. Today advanced chips contain a billion transistors.

The vast improvements from the abacus to the Pentium chip exemplify the promise of nanotechnology.

Gordon Moore, who was the chief research officer at the Intel Corp, predicted this trend early in the semiconductor development. It has well described the path to the billion-transistor chip of area about $1\,\text{cm} \times 1\,\text{cm}$. Successive steps in miniaturization, foreseen by Moore, have always been dependent on market demand. Interplay between applications and sales were essential. The successive improvements required new instrumentation and capital investment. So it was, and remains, a cooperative phenomenon between what is possible and what is profitable. The end result, however, has been beneficial. The present situation is that the industry annually produces approximately 10^{18} (1 billion billions) field effect transistors (FETs) and the cost per unit has fallen to the range of nanodollar [1].

Even though this is not universally recognized (but see Ref. [1] of Chapter 1), the transistor chip phenomenon is nanotechnology, on any rational basis. Each transistor chip contains eight or more separate layers engineered on the scale 10–100 nm into the surface of the silicon chip.

Nanotechnology has been defined by the National Science Foundation as design and controlled production of devices with at least one dimension in the range of 1–100 nm.

The devices that are engineered into the surface of the silicon are identical, produced by step-and-repeat photolithography, which we will come back to later. In the end, a large mask is created, which can pattern thousands of identical transistors simultaneously. We will come back later to discuss some of the techniques that have been used to make this happen within the engineering community. The field is based upon the original invention of the transistor, recognized by the Nobel Prize in Physics to John Bardeen, Walter Brattain, and William Shockley, who were, respectively a physicist, a chemist, and an electrical engineer.

In the case of the transistor, the vastly increased number of devices available on one chip has made a great increase in functionality on a given chip. This is the primary benefit. There are more devices in the same area, and at lower cost. The secondary benefit is that the individual devices work faster. To get to this condition, as we have said, has required repeated large investments for improved equipment. The investments that are needed are unfortunately getting bigger, accelerating, because the state of the art is so far advanced. The cost of a new semiconductor fabrication facility is billions of dollars, which means that only the largest firms can compete.[5] The advance, with reduced size, in the functioning of the individual devices, is a topic we will return to, but it fits the general picture that smaller things are more rapid in their operation. So "smaller is better," smaller in the first instance means more, and in the second instance means that the devices act faster.

A major issue limiting further advance is the large amount of power to run the chip. This is a problem that may not be soluble without a major change, from a semiconductor device, for example, to a superconductor device. (We will talk about this in Chapter 13.) The power dissipated in the semiconductor device is the product of a voltage times a current, in a simple picture. The voltage required to turn on the device is in the vicinity of a volt and can't be reduced. The current when the device is on has apparently fallen, since the power per chip has remained in the vicinity of 100 W as the count of transistors has increased. Most chips do not require more than 100 W to run, and although this is more than the power required originally in the smaller chips of G. Moore's day, it is a workable power only because the single chip will support an entire laptop or desktop computer. So the increased functionality justifies the large power that the chip needs. Arguably, Moore's law and the vastly improved chip, always driven by demand for better performance, has led to the present situation now enhanced with the additional factor optical fibers and Internet communication, by the ability to store and quickly access large amounts of information in *cloud computing*.

What is *cloud computing*? The demand for computing has risen to the point that everybody has to have a computer. Larger data files in pictures and videos have required faster computers with more storage space, so the trend is toward everyone needing a supercomputer! In order to reduce costs with the availability of the Internet, it makes sense to provide central facilities where large amounts of data can be stored and also where rapid calculations can be performed. So the idea is that everybody does not have to *own* a supercomputer; they can log into the cloud computer, which is perhaps a supercomputer on the Internet, and ask that computer to do the task. The demand for large amounts of storage has been stimulated by the change in photography from traditional film to digital camera, whose output is immediately a data file that can be sent on the Internet. So there has been, as everybody knows, a great proliferation (now to 800 M) of users on Facebook, most with images, or even videos, that their friends can see. Such images are typical of data stored on cloud computers, accessible via the Internet from wherever the user may be. So, this is an efficiency and a convenience and in sum it has allowed a tremendous improvement in the ability of people to communicate. We can consider cloud computing and social networks as developments benefiting from the basic notion that "smaller is better." (To

2.3
Miniaturization: Esaki's Tunneling Diode, 1-TB Magnetic Disk "Read" Heads

The notion that "smaller is better" is central to nanotechnology, and we have given perhaps the leading example, which is the rise of Moore's law in computing capacity in the laptop computer driven by a single silicon chip.

Within the category of computer technology, there are further subtopics illustrating how and why "smaller is better."

One of these is the magnetic disk storage technology. The hard disk on your computer may hold 100 GB, which is a tremendous amount of data, and can be accessed on a millisecond timescale. The PC hard disk, and similar and larger disks that are the central part of a cloud computing facility, is a dense data storage medium.

Disk memory has recently followed an accelerated form of Moore's evolution. The most recent substantial advances made both the storage elements and the reading devices smaller (see Figure 2.1). We address these in turn.

The storage elements are ferromagnetic domains, like tiny bar magnets,[6] deposited and patterned into the surface of the hard disk. These storage domains are now smaller and their orientation has been changed to make them perpendicular to the disk surface, to require less space.

The basic unit of storage (Figure 2.1) is a nanometer-scale "bar magnet" that has its magnet field or magnetization *parallel* or *antiparallel* to its longest dimension. The domain magnetic field direction can be switched, by change of the orientations of electron spins inside the domain. This is an element of nanophysics explained in Chapter 11.

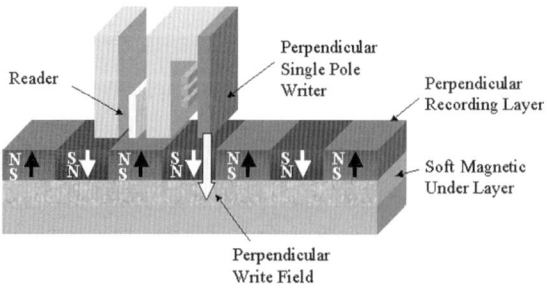

Figure 2.1 Schematic diagram of read–write head for perpendicular magnetic bits on modern terabyte hard-disk drive, cost $100, capacity 1/10 of Library of Congress. This will be described in Chapter 11 [3].

2.3 Miniaturization: Esaki's Tunneling Diode, 1-TB Magnetic Disk "Read" Heads

The smaller-size reduced area of domain has been enabled by a change in the "read head." The reader is now able to access a single domain whose cross section is about 100 nm × 100 nm. This means that the reading device, in order to see only one domain, has to be small, and a recent advance in nanotechnology has made the "read head" much smaller, as shown in Figure 2.1.

The present *reader* (as labeled in Figure 2.1), called TMR for tunneling magnetoresistance,[7] uses a magnetic tunnel junction (MTJ). The MTJ is a three-layer sandwich, two "electrodes" separated by a "tunnel barrier." The electrodes are made of ferromagnetic metals, like iron, nickel, or cobalt, and they are so small that they make up a single "magnetic domain."

The physics of the MTJ reading device (Chapters 5 and 11) is that the *relative orientation of the two ferromagnetic electrode domains controls the tunneling resistance through the insulator.* (This electrode-domain orientation follows the magnetic field emanating from the data domain in the disk surface.) If the MTJ domains are parallel, current flows, but is much reduced if the domains are antiparallel. This is the basis of the reader. The beauty is that the tunnel junction device can be tiny, small enough to have a single magnetic domain, whose size is 100 nm × 100 nm.

After costly research and development, MTJ devices are now cheap. They are made by the millions using technology (see Figure 2.2) similar to that

Figure 2.2 Highly magnified scanning electron microscope (SEM) view of IBM's six-level copper interconnect technology in an integrated circuit chip. The channel lengths in this chip are 120 nm. (Photograph courtesy of IBM Corporation, 1997.) [4]

leading to Moore's law [5]. This change of technology from the original disk reader, a "pickup coil,"[8] based on quantum physics of magnetism (recognized by the Nobel Prizes[9]), leading to the quantum tunnel effect in the MTJ, has had a large economic impact.[10]

The sensing feature of MTJ is that one film has an easily switchable magnetization, is called a soft ferromagnet, which allows a 10 mT or smaller magnetic field to flip or reverse its orientation. The electrode is so small that it has only one domain, and linear, so only two choices, up or down along sheet, tunnel junction electrode. The second film in the MTJ is a "hard" magnet whose magnetization will not change, like the one you buy in hardware stores. The MTJ device encounters field from the magnetic domain on the disk surface, will have its soft magnetic layer oriented to be parallel to the field from the surface, thus it reads the field from the surface. MTJ conducts best when orientations are parallel, and turns off in antiparallel configuration. Accordingly, MR is defined as $MR = (R_{AP} - R_P)/R_P$, with R_{ap} and R_p, respectively, the resistance with the electrode magnetizations being antiparallel and parallel.

The MTJ reader is an example of "smaller is better," but in this case smaller has been enabled by being different. The route to "smaller" has gone through nanophysics, which offers a different detection method, easier to miniaturize.[10]

2.4
Accelerometers and Semiconductor Lasers

Tiny accelerometers are part of the nanoelectronics revolution. This is a sensor that detects acceleration, and gravity is one form of acceleration. Nanoelectronic semiconductor chip accelerometers were first used widely in airbag sensors for cars. (They are now used to maintain balance in Segway two-wheel transporters and in iPod Nano devices to reorient the display.) The air bags will inflate if the sensor detects a strong acceleration. The sensor in these devices is a cantilever, essentially a spring with a mass attached to the end. In practice, the cantilever itself has a distributed mass, and for purposes of calculations one can replace distributed mass by an imaginary point mass at the midpoint. Cantilevers of tiny size, as small as 50-μm length, with spring constant 1 N/m and resonant frequencies of 200 kHz can be mass fabricated in silicon technology from single crystals of silicon (as we will discuss in Chapter 11) in connection with the IBM Millipede mass storage device.

In the airbag accelerometer, the cantilever mount is rigidly attached to the frame of the car. Supposing the support of the cantilever suddenly jumps upward, the mass of the cantilever resists motion (inertia), so the spring deflects. This amounts to a downward deflection in its own frame of reference attached to the car, and can be detected. (One way would be in electrostatics, by a changed capacitance.) Such an accelerometer can be extended to provide a 3D detection of acceleration. Sensitive accelerometers recording vector acceleration over time in 3D can record the trajectory. This could tell a spacecraft where it is, on the way to the moon or to Mars, in a region where the global positioning system (GPS) satellites are not available!

Another example of "smaller is better" is provided by the class of filters that are called Nanopore filters. The nano is built into the trade name of this class of products! These filters have pores, openings, which are variably available on the nanometer scale. These filters are based on nuclear technology. The pores, which are small and oriented perpendicular to the polymer film of the filter, are produced by exposing the polymer to a flux of energetic charged particles. These might be protons, or alpha particles, anything with electric charge and enough energy to fly right through the polymer film. This produces in the film a track of damage and if the polymer is then suitably chemically etched, the damaged material in the tracks can be removed, leaving holes or pores that are straight and parallel and perpendicular to the film. These filters are strong, made in strong plastic films, and they can be used, for example, in water purification. The size range of the filters is such that water will go through but even bacteria will be too big to go through. So, a high-class water supply for a hiker or a world traveler who doesn't trust the water to drink can use a Nanopore filter setup to be safe. The filters are inert and not easily destroyed by exposure to dirty water, and if they get plugged by silt, they can be washed off.

Other examples of "smaller is better" are the platelets of silver halide, silver bromide, and silver chloride, which form the basis for conventional film photography. The recording units in the film are tiny controllably produced silver halide crystals that are sensitized by absorbing light. The sensitization makes the illuminated crystal susceptible to being turned into a tiny chunk of silver, if put in a developer solution. So the light image is turned into an image of silver particles, embodied into the negative, which is thus very durable. The advantage of silver halide photography is that the basic image is metallic silver, and likely to survive for thousands of years, almost like words written into stone, or etchings on ceramic pots from centuries ago. One of the disadvantages of the silicon technology is that means for retrieving information from earlier versions of the technology sometimes disappear! If you have your life history written on the old-style floppy disk, your heirs may not be able to recover that information. You cannot easily buy a computer nowadays to read the old-style floppy disk. Silver halide does not suffer from this drawback; it is a permanent record, not dependent upon a fancy machine to read it.

These advances in storage nanotechnology have come not only from making some initial device smaller, but also, in an important part, from pivotal discoveries along the way, to make entirely new sensor devices. This aspect of nanotechnology we think is important and justifies a broadening of the definition of nanotechnology to include devices and machines whose operation depends upon engineering knowledge of nanophysical laws. In the case of the disk reader this means understanding at the engineering level of how electron spins interact, and the electron spins up or down are nanophysics and the disk reader is a magnetic field detector whose operation depends upon a device that can count the number of spins of *one orientation*, that tunnel across the barrier. So this device depends on the nanophysics, not only of spin ½, but also depends on a separate basic nanophysical behavior, the tunneling of electrons through a barrier of an MTJ.

Other examples of improvements that have come along with miniaturization in the semiconductor industry include the *injection or junction laser*, a specific form of the P-N junction, which we will describe later, and the optical fiber. The injection laser provides light in the wavelength range of 1–1.5 μm to illuminate optical fibers that transmit information worldwide. One can argue that the fibers themselves are a product of miniaturization. The best fibers are now small enough to carry a single optical mode. The fiber itself may be larger than nanometers, with a cylindrical core of 4-μm radius of silica or glass with index of refraction n in the range of 1.5–3, surrounded by a larger coaxial cylinder of glass whose index is slightly smaller. The index of refraction $n = c/c'$ where c' is the speed of light in the medium, and c is light speed in vacuum, 3×10^8 m/s. The light concentrates to propagate with low loss by the single-mode aspect, which keeps the light inside away from the outer dirty surfaces. The light signal along the fiber can be amplified by another nanophysical device, the erbium-doped fiber amplifier, similar in its operation to a laser (light amplification by stimulated emission).[11]

The physics of *stimulated emission* was discovered by Einstein in 1905 and forms the basis of all laser technology.

2.5
Nanophysics-Based Technology: Medical Imaging, Atomic Clock, Sensors, Quantum Computers

If the definition of *nanotechnology* is enlarged to include devices that operate on nanophysical principles, then the inventory is quite large and would include nuclear reactors and atomic clocks, mentioned in Chapter 1. The operating parts in the fission reactor are nuclei of ^{235}U, which break up into smaller nuclei in fission events. In the generalized view of nanotechnology we can include the atomic clock, which depends on knowledge of electron and nuclear spins and their interaction inside in an atom and their interaction with microwave photons. The change of the spin configuration in ^{133}Cs leads to microwave radiation at 9.1 GHz. The *second* is defined as 9.1 billion oscillations of this spin system.

Another by-product making things smaller, through Moore's law evolution of semiconductor electronics, is the improvement of semiconductor devices to count 9.1 billion counts per second. To make use of a clock that ticks at 9.1 GHz you need a device to record those ticks. The tick of the clock is the spin transition. Smaller transistors, operating faster, actually now allow counters that will work up to 200 GHz.

We have summarized in Chapter 2 the essential improvement in computing that has come with making transistors smaller, which is called Moore's law. The count of transistors per chip has gone from about 30 in 1965 to a billion today. The speed of operation of the devices has increased as a consequence of making them smaller. Devices now work at least at 2 GHz and some semiconductor counters operate up to 200 GHZ. Magnetic storage in the hard-disk drive has benefited from miniaturization and also by the discovery, as a part of the miniaturization process, of a new magnetic field sensor, the MTJ. This is an unexpected discovery, a new operating principle based on nanophysics. Other technological advances based on knowledge of nanophysics include the atomic clock and the injection laser.

In Chapter 3 we turn to "scaling," the rules needed to engineer and manufacture devices of smaller sizes.

References

1 Pinto, M.R. (2007) in *Future Trends in Microelectronics: Up the Nano Creek* (eds S. Luryi, J. Xu, and A. Zaslavsky), John Wiley & Sons, Inc., Hoboken, NJ, p. 154.
2 Havelock, R.G. (2011) *Acceleration: The Forces Driving Human Progress*, Prometheus Books, Amherst, NY.
3 Toshiba Corp (2004) Press release dated 14 December 2004, Announcing "the world's first hard disk drives based on perpendicular recording, a breakthrough technology."
4 Edelstein, D. (2011) Copper interconnects: the evolution of microprocessors, http://www.ibm.com/ibm100/us/en/icons/copperchip/ (accessed 5 December 2011).
5 Campbell, S.A. (2001) *The Science and Engineering of Microelectronic Fabrication*, 2nd edn, Oxford University Press, New York.

3
Systematics of Scaling Things Down: $L = 1\,\text{m} \to 1\,\text{nm}$

To get oriented, let us recall from Chapter 2 that the nanometer is one billionth part of a meter.

The smallest lateral dimension in the present transistors in computers is about 45 nm, thus about 450 atoms wide. A state-of-the-art computer chip contains a billion transistors of this sort. An interesting fact that suggests the importance of nanotechnology is the production of *field effect transistors* (FETs), which are the central device in computers and information technology. *One billion billions* of these devices are produced per year. We will return later to explain these devices.

3.1
One-Dimensional and Three-Dimensional Scaling

A primary working tool of the nanotechnologist is facility in scaling the magnitudes of various properties of interest, as the length scale L shrinks, e.g., from 1 mm to 1 nm.

Clearly, the number of atoms in a device scales as L^3, which means a constant number of atoms per unit volume. If a transistor on a micrometer scale contains 10^{12} atoms, then on a nanometer scale, $L'/L = 10^{-3}$, it will contain 1000 atoms, likely too few to preserve its function!

Normally, we will think of *scaling* as an isotropic scale reduction in three dimensions. So a scaled cube will remain cubic. However, *scaling* can be thought of usefully when applied only to one or two dimensions, scaling a cube to a two-dimensional (2D) sheet of thickness "a" or to a one-dimensional (1D) tube or "nanowire" of cross-sectional area a^2. The term "zero-dimensional" or "quantum dot" is used to describe an object small in all three dimensions, having volume a^3.

Figure 3.1 Transmission electron micrograph (TEM) image of one 5-nm CdSe quantum dot particle (courtesy of Andreas Kadavanich and Quantum Dot Corporation).

A useful property of engineering materials, including metals and semiconductors, is the *reliability or continuity of concepts and approximate parameter values down to about* $L = 10\,nm$ *(100 atoms on a side)*. This forms the "classical" range of scaling, from 1 mm down to perhaps 10 nm. It is related to the stability (constancy) of the basic microscopic properties of condensed matter (conventional building and engineering materials) almost down to the scale L of 10 nm or 100 atoms in line, or a million atoms per cube. Amazing but true, and the idea can be seen in Figure 3.1.

Even though this particle is only about 8 nm across, you can count the atoms; there are enough atoms in this particle so that locally its properties are quite similar to those in a much larger sample! This makes scaling to small sizes an engineering exercise, we can accurately predict what will happen because we know that the basic underlying parameters do not change very much. *(The following section is perhaps engineering oriented and certainly numerical. It might be skipped by a casual reader. It would be essential reading for a professional planning to make a new career in nanotechnology, or for any person who would like to learn what is behind the "clock" in his PC computer. And certainly see Figure 3.5.)*

Typical microscopic properties of condensed matter are the interatomic spacing (the "lattice constant"), the mass density ρ (kg/m^3), the bulk speed of sound v_s, Young's modulus Y, and the shear or rigidity modulus S (both have units GPa, or gigapascals [10^9 Pa] as we will explain). The pascal is a pressure of 1 N/m^2, and, for orientation, 1 atm is 101 kPa, and the cohesive energy U_o (Figure 3.2).

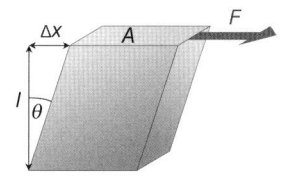

Figure 3.2 Definition of shear modulus $S = (F/A)/(\Delta x/l)$, whose value is usually expressed in GPa. Young's modulus Y is similar, but the elongation (compression) and the force are both in the vertical direction, say z. These moduli express the essential stiffness of a material and have units of pressure because the *strain* $(\Delta x/l)$ or $(\Delta z/l)$ are dimensionless. These parameters do not change much as the size of a sample is reduced down toward 100 atoms on a side [1].

3.2
Examples of Scaling: Clocks, Tuning Forks, Quartz Watches, Carbon Nanotubes

The basic idea of scaling is simply illustrated in the case of the pendulum clock. The period of a grandfather clock is 2 s. You get one click each time the pendulum reverses direction and the escapement, which couples energy from the weights, into the swinging motion, makes a click. The period of a pendulum T comes from Newton's laws of classical physics

$$T = 2\pi \, (l/g)^{1/2} \tag{3.1}$$

where l is the length of the pendulum and g is the local acceleration of gravity, approximately 9.8 m/s². The grandfather clock has a fixed length l of 1 m for a period of $T = 2$ s. If we want to make a faster pendulum clock we can scale down the length l. So an easy exercise is what is l to have a pendulum oscillator at A on the musical scale, A = 440 Hz. Since the dependence of f is on the square root of l, the new l will be $(0.5/440)^2 = 1.29 \times 10^{-6}$. So the required length L' is 1.29 μm, an impractical value. This is an exercise in scaling in one dimension, only "l" is changed.

Now, a pendulum clock is useful really only in a fixed location, it won't operate upside down, nor if the local acceleration of gravity is changed, as for example, in a falling elevator or an airplane pulling out of a dive. A more practical oscillator is one based on a spring and a mass instead of a pendulum. This is like the cantilever we mentioned in Chapter 2 as the operating part of the airbag accelerometer, and also of the IBM Millipede dense storage device that we will come back to in Chapter 11. The frequency of a mass on a spring of constant K is

$$f = (1/2\pi)(K/m)^{1/2} \tag{3.2}$$

where K is the spring constant, with units of N/m. (One Newton is about 0.22 of lb force.) The period of a clock based on a spring oscillator[1] is just $T = 1/f$, as defined above. A linear (Hooke's law) spring is one whose force is $-Kx$, where x is the extension of the spring. This equation strictly describes a mass m attached to a massless spring described by the spring constant K. However, the formula is adaptable to the real-world cantilever, an elastic bar like a diving board, described by a length L, width w and thickness t.

If a cantilever, such as a diving board, is clamped at one end, the resonant frequency is given by a formula still in the form of $f = (1/2\pi)(K^*/m^*)^{1/2}$, but the K^* and m^* are somewhat altered by the distributed geometry of the oscillator. The effective spring constant K^* involves the material stiffness parameter, Y (Young's modulus, defined as pressure divided by fractional distortion). The mass is now distributed through the volume of the cantilever, and enters in the form of the material density ρ in units of kg/m³. A working formula for the frequency of oscillation for a cantilever clamped at one end is

$$f = 0.162\,(t/L^2)(Y/\rho)^{1/2} \tag{3.3}$$

Here, L is the length and t is the thickness of the rectangular bar, and notice the width does not enter. $(Y/\rho)^{1/2}$ is a materials parameter, the ratio of the stiffness to the mass density, which enters as the square root, and may not change much among similar materials. This formula is still essentially $\sqrt{(K/m)}$, but it breaks out the stiffness, mass density, the length, and the thickness of the rectangular bar (like a diving board) clamped at one end.[2] *The essential scaling parameter is t/L^2.*

This formula is typical, in that it shows the separate roles of the material properties and the dimensions. The material parameters stiffness and mass density are constant if we change only the dimensions, but we have to take account of these parameters when comparing one material to another. Also, the *scaling parameter t/L^2 reveals that the frequency varies as $1/L^2$ if we hold t constant, but scales in a 3D sense as $1/L$, if we assume length and thickness are uniformly changed in the same proportion (isotropic scaling).*[3]

The musical tuning fork, with two prongs, actually is really a cantilever, the two prongs act independently. The most common tuning fork is at musical pitch $A = 440\,\mathrm{Hz}$. Let's take a concrete example and scale it to smaller size.

So using our formula (3.3) to model a steel 440-Hz tuning fork, with dimensions $L = 6.6\,\mathrm{cm}$, $t = 0.28\,\mathrm{cm}$, nominal $Y = 200\,\mathrm{GPa}$, and density $\rho = 7850\,\mathrm{kg/m^3}$ gives 534 Hz, 20% bigger than the measured frequency $A = 440\,\mathrm{Hz}$. This is not too bad agreement for a back-of-the-envelope calculation.

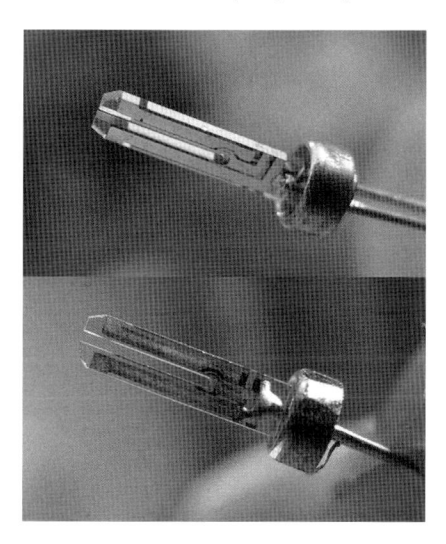

3.2 Examples of Scaling: Clocks, Tuning Forks, Quartz Watches, Carbon Nanotubes

Figure 3.3 The Bulova quartz cantilever oscillator, frequency 32.8 kHz [2].

An important quartz oscillator is the Bulova quartz watch cantilever (Figure 3.3), which oscillates at 32.8 kHz. Let's see if we can scale what we have learned from the 440-Hz tuning fork to predict the Bulova quartz oscillator frequency.

The scaling is in two parts, the dimensions and in the materials properties. The Bulova quartz oscillator dimensions are given as $L = 3$ to $4\,\text{mm}$, $t = 0.3\,\text{mm}$, $Y = 100\,\text{GPa}$, and $\rho = 2634\,\text{kg/m}^3$. Let's do our calculation by solving for the length L' of the Bulova oscillator, since it was not given precisely, using the formula (3.3) and the changes of dimensions and of material properties. Working only by the ratios defined in Eq. (3.3), to find L' we write the ratio:

$$f'/f = 32\,800/440 = (0.3/2.8)(66/L')^2 [100/200 \times 7850/2634]^{1/2} \quad (3.3a)$$

The answer from this is L' (the predicted length of the Bulova cantilever, scaling from the steel tuning fork), is 2.75 mm, vs. stated 3–4 mm. This is again not perfect but it is not too far off.

Another example of a cantilever is a silicon nanomachined cantilever [3] that is anchored at both ends (which multiplies the formula (3.3) by a factor 8.45) has a frequency 0.4 GHz at length $L = 2\,\mu\text{m}$ with width $t = 66\,\text{nm}$. Neglecting the material parameter change from quartz to silicon let's scale from 32.8 kHz to the new dimensions,

$$f = [0.162\,(t/L^2)(Y/\rho)^{1/2}] \times 8.45 \text{ (doubly anchored beam)} \quad (3.4)$$

so $f' = 32.8\text{ kHz} \times 8.45 \times (66\,\text{nm}/0.3\,\text{mm})(3.5\,\text{mm}/2\,\mu\text{m})^2 \quad (3.4a)$

which gives $f' = 186.8$ MHz. This answer is to be compared [3] with 400 MHz, again it is not precise, but not too far off! We have now scaled by a factor 0.43 million from the tuning fork at 440 Hz, and we have gotten here using the same basic formula (3.3, 3.4) by scaling the dimensions and the material parameter.

A device that has been proposed [4] as a high-density random access memory is a crossbar array of carbon nanotubes, which have a diameter of 1.37 nm. The design includes 20 nm spacing of pillars that support the nanotube at each end. So this geometry is like the previous one, it is a doubly clamped beam of length 20 nm and radius 0.69 nm, described by Eq. (3.4). The density of diamond is 3.52 g cm^{-3} and Young's modulus Y is 10^{12}. This value is $f = 29$ GHz if we assume the density is the same as diamond for the carbon nanotube. The authors of the paper predict 100 GHz. The value of L should be smaller, because the 20 nm was between supports, and we have neglected the width of the supports. So again this is not too far off.

An important clock in our daily existence is that in the PC or laptop computer. My computer says it operates at 1.8 GHz. How is that frequency established? It is established, again, by a quartz oscillator. However, the quartz oscillator in this case is a lot simpler. It is just a slab of quartz with metallization on each side. (The same geometry is also used as a deposit thickness monitor in vacuum practice, and in this application it is called a "quartz crystal microbalance." The frequency of oscillation slightly changes with deposit, and this change can be accurately measured.) If we put a voltage across the device, which is like a capacitor whose dielectric is quartz, it will slightly distort, because quartz is a piezoelectric material whose dimensions are slightly affected by an electric field.

Let's think of a slab of quartz of thickness t, of area "A," as in Figure 3.4, ignoring the biological aspect. The mass is $\rho t A$, the product of mass density ρ, thickness t, and area A. If we compress or stretch the quartz in the direction of t, we can use Young's modulus Y to get the force F: From the definition of Young's modulus we have

$$Y = (F/A)(t/\Delta t), \text{ so } K = F/\Delta t = AY/t \qquad (3.5)$$

This is a one-dimensional oscillator, and, since only half of the mass moves (the midpoint is motionless), the frequency is

$$f = (1/2\pi)(2AY/tA\rho t)^{1/2} = (1/2\pi)(1/t)(2Y/\rho)^{1/2} \qquad (3.6)$$

If we make the quartz crystal 5.0 μm thick we get

$$f = (1/2\pi)(1/5.0\times10^{-6})(2\times100\times10^{9}/2634)^{1/2} = 0.277 \text{ GHz} \qquad (3.6a)$$

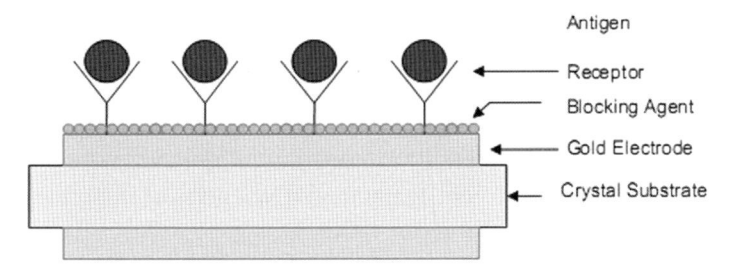

3.2 Examples of Scaling: Clocks, Tuning Forks, Quartz Watches, Carbon Nanotubes

Figure 3.4 Quartz crystal microbalance (basically the same as the PC clock) here set up to detect antigens, on top, which will add small mass and slightly reduce the frequency. In usual practice, the oscillation is transverse, a shear distortion, as suggested by Figure 3.1. The text describes a variant in which the oscillation will be vertical, in the z-direction, to involve Young's modulus Y rather than the shear modulus S. The equations are valid in either case [5].

This is then multiplied in circuitry to get to the 1.8 GHz which is the typical PC clock frequency. It appears that this frequency could be increased by choosing the crystal thickness less than 5 μm thick.[4] With electrodes (not included in our estimated frequency) this is called a "quartz crystal microbalance" and the change of frequency with deposited mass is used as a thickness monitor.

A voltage will start the oscillation with a frequency determined by the small thickness t of the quartz layer. Quartz is desirable because it is loss-free, which corresponds to a very high Q value for the oscillator. The Q is defined as the frequency divided by the frequency width of the oscillation, it also measures by how many cycles the oscillation will last if the power is turned off. The Q for a quartz oscillator can be in the range of 10^5. This quartz oscillator plays the same role as an LC circuit, that is, as a frequency determining element, and it will be kept in oscillation by an amplifier that gives positive feedback. This will not need much energy to run because of the high Q of the oscillator. This kind of oscillator can be made to run as high as 0.2 GHz by making the quartz layer very thin. To get from there to the clock frequency 1.8 GHz, which is done in all PC computers, is based upon multiplying this frequency.

The current state of the art in timekeeping is the atomic clock, and a standard model is the Hewlett Packard Model 5071A that is used in the global positioning system (GPS). The atomic clock is often used to stabilize a quartz oscillator, or a quartz oscillator frequency that has been multiplied in semiconductor circuitry.

The limiting frequency is mechanical oscillators of the type we have been considering are actually molecules, such as H_2 or CO. The same formula

$f = (1/2\pi)(K/m)^{1/2}$ applies to these molecular oscillators, where the frequencies approach 10^{14} Hz and spring constant K physically arises in the covalent bond between the two atoms in the molecule. The covalent bond equivalent spring constant values are in the range of 1000–2000 N/m. For example, the frequency is 64.3 THz = 6.4×10^{13} Hz for CO and the spring constant for the CO molecule is about 1860 N/m. The vibrational frequencies of molecules are known from absorption of light in the infrared range.

The hydrogen molecule vibration frequency is 1.32×10^{14} Hz and its equilibrium spacing is 0.074 nm. If we return to our formula for the quartz oscillator at 0.277 GHz and 5-μm spacing, and scale that frequency to a thickness $t = 0.074$ nm, we find 1.9×10^{13} Hz. We should not expect this to be accurate, because in going from a shear-motion oscillation in crystalline quartz to simply two protons on a "spring" formed by a single covalent bond the rules governing the oscillation have substantially changed! But clearly the trend to higher frequency at smaller length L is maintained.

The scaling examples we have discussed are summarized in the plot of Figure 3.5. Log f vs. log $1/L$ should have a slope 2 for the 1D scaling that we

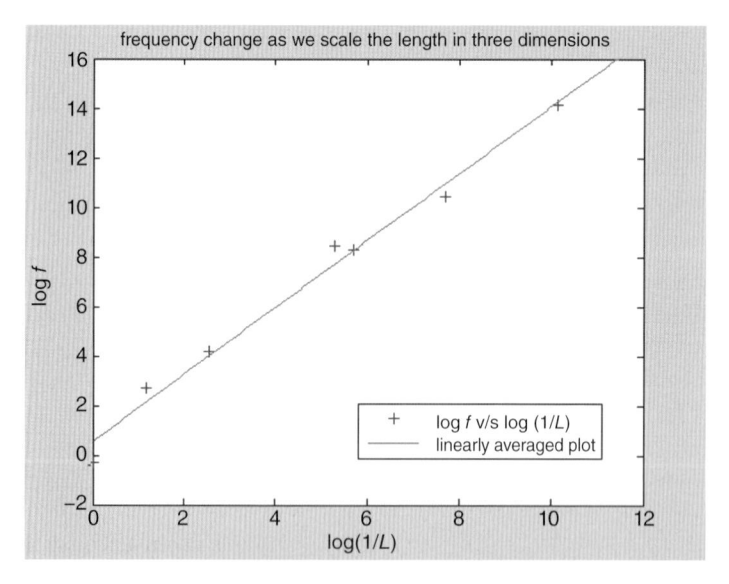

Figure 3.5 Empirical evidence that smaller objects exhibit higher vibration frequencies. In this plot the lengths L are expressed in meters. We see a frequency range of 14 orders of magnitude from the grandfather clock ($L = 1$ m, $f = 0.5$ Hz) to the hydrogen molecule at f about 10^{14} Hz. Data are summarized in Note N5. Empirically, the data represented by the line follow a rule $f \propto (1/L)^n$ with $n = 1.32$. In the text Eq. (3.3) (see Note N3) explains that $n = 2$ is expected for scaling in one dimension, while $n = 1$ is expected for isotropic scaling. The surveyed oscillators in nature behave in an intermediate fashion.

have mostly discussed as in Eq. (3.3). A slope of 1 would apply to isotropic scaling as described in Note N3 and in the text. The values used to construct this plot are shown in N5.[5] (**More about scaling:** this section is more technical and could be skipped by some readers.)

In this section we adopt $\omega = 2\pi f$ as the frequency, because it simplifies many formulas.

Insight into the typical scaling of other kinetic parameters such as velocity, acceleration, energy density, and power density can be understood by further consideration of such simple harmonic oscillators, represented by the cantilever. (In what follows we assume isotropic scaling, all three dimensions are changed simultaneously.)

A reasonable quantity to hold constant under scaling is the strain, x_{max}/L, where x_{max} is the amplitude of the motion and L is length. So, the peak velocity v of the mass $v_{max} = \omega\, x_{max}$ that is then constant under scaling: $v \propto L^0$, since $\omega \propto L^{-1}$. Similarly, the maximum acceleration "a" is $a_{max} = \omega^2\, x_{max}$, which then scales as $a \propto L^{-1}$. (The same conclusion can be reached by thinking of a mass in circular motion. The centripetal acceleration is $a = v^2/r$, where r is the radius of the circular motion of constant speed v.) Thus, for the oscillator under isotropic scaling the total energy $U = \tfrac{1}{2} K x_{max}^2$ scales as L^3.

In simple harmonic motion, the energy resides entirely in the spring when $x = x_{max}$, but has completely turned into kinetic energy at $x = 0$, a time $T/4$ later. The spring then has done work U in a time $1/\omega$, so the power $P = dU/dt$ produced by the spring is $\alpha\, \omega U$, which thus scales as L^2. Finally, the power per unit volume (power density) scales as L^{-1}. *The power density strongly increases at small sizes.* These conclusions are generally valid as scaling relations.

3.3
Scaling Relations Illustrated by Simple Circuit Elements

A parallel plate capacitor of area A and spacing t gives $C = \varepsilon_o A/t$, which under isotropic scaling varies as L. The electric field in a charged capacitor is $E = \sigma/\varepsilon_o$, where σ is the charge density. This quantity is taken as constant under scaling, so E is also constant. The energy stored in the charged capacitor $U = Q^2/2C = (1/2)\,\varepsilon_o E^2 At$, where "$At$" is the volume of the capacitor. Thus, U scales as L^3. If a capacitor is discharged through a resistor R, the time constant is $\tau = RC$. Since the resistance $R = \rho \ell /A$, where ρ is the resistivity, ℓ the length, and A the constant cross section of the device, we see that R scales as L^{-1}. Thus, the resistive time constant RC is constant (scales as L^0). The resistive electrical power produced in the discharge is $dU/dt = U/RC$, and thus scales as L^3. The corresponding resistive power density is therefore constant under scale changes.

For an LC circuit the charge on the capacitor $Q = Q(0)\cos[(C/L)^{1/2}t]$. The radian resonant frequency $\omega_{LC} = (C/L)^{1/2}$ thus scales as L^0. That is to say, the frequency does not change under isotropic scaling.

3.4
Viscous Forces for Small Particles in Fluid Media

Viscous forces become dominant for small particles in fluid media. The motion of a mass in a fluid, such as air or water, eventually changes from inertial to diffusive as the mass of the moving object is reduced. Newton's laws (inertial) are a good starting point for the motions of artillery shells and baseballs, even though these masses move through a viscous medium, the atmosphere. The first corrections for air resistance are usually velocity-dependent drag forces. A completely different approach has to taken for the motion of a falling leaf or for the motion of a microscopic mass in air or in water.

The most relevant property of the medium is the viscosity η, defined in terms of the force $F = \eta v A/z$ necessary to move a flat surface of area A parallel to an extended surface at a spacing z and relative velocity v in the medium in question. The unit of viscosity η is the pascal second (1 Pa is a pressure of 1 N/m²). The viscosity of air is about 0.018×10^{-3} Pa s, while the value for water is about 1.8×10^{-3} Pa s. The traditional unit of viscosity, the Poise, is 0.1 Pa s in magnitude.

The force needed to move a sphere of radius R at a velocity v through a viscous medium is given by Stokes' law,

$$F = 6\pi\eta Rv \quad (3.7)$$

This is valid only for very small particles and small velocities, under conditions of streamline flow.[6]

The fall, under the acceleration of gravity g, of a tiny particle of mass m in this regime is described, following Stokes' law, by a limiting velocity obtained by setting F (from Eq. 3.7) equal to mg. This gives

$$v = mg/6\pi\eta R \quad (3.8)$$

As an example, a particle of 10-μm radius and density 2000 kg/m³ falls in air at about 23 mm/s, while a 15-nm particle of density 500 kg/m³ will fall in air at about 13 nm/s. In the latter case, one would expect random jostling forces $f(t)$ on the particle by impacts with individual air molecules (Brownian motion) to be present as well as the slow average motion. Newton's laws of motion as applied to the motion of artillery shells are not useful in such cases, nor for any cases of cells or bacteria in fluid media.

3.5
What about Scaling Airplanes and Birds to Small Sizes?

In the broader but related topic of flying in air, a qualitative transition in behavior is observed in the vicinity of 1-mm wingspan. Lift forces from smooth flow over airfoil surfaces, which derive from Bernoulli's principle, become small as the scale is reduced. The flight of the bumblebee is not aerodynamically possible, we are told, and the same conclusion applies to smaller flying insects such as mosquitos and gnats. In these cases the action of the wing is more like the action of an oar as it is forced against the relatively immovable water. The reaction force against moving the viscous and massive medium is the force that moves the rowboat and also the force that lifts the bumblebee.

No tiny airplane can glide, another consequence of classical scaling. A tiny airplane will simply fall, reaching a terminal velocity that becomes smaller as its size is reduced.

Another topic of scaling important in nanotechnology has been the scaling process that has been applied to FETs in the Moore's law miniaturization. This scaling, which has led to the 45-nm-scale node wiring dimension is nominally based on the idea of keeping the *electric field in the devices constant*. The scaling of the silicon devices has been essentially isotropic, reducing the size in x, y, and z. The reduction in z, the thickness of the layers built into the silicon as we will see, has had to be corrected when the SiO_2 traditional oxide became so thin that electron tunneling produced noticeable leakage between the gate and the channel. As we will see later this has been cured by the method of atomic layer epitaxy and the growth of HfO_2 rather than SiO_2, as the gate insulator. The question remains as to how small the devices can be made.

In this chapter we have discussed the numerics of making devices smaller. The basic assumption is that the material properties such as density and stiffness, Young's modulus, are not changed but that the changes come only from the change in dimensions. We discussed changes in the resonant frequency of spring oscillators from the tuning fork at 440 Hz down to the Bulova watch quartz oscillator at 32.8 kHz, to some nanomachined silicon bars, doubly clamped beams that were observed to oscillate near 400 MHz, to a carbon nanotube of length 20 nm clamped at both ends that we estimated as 29 GHz. The clock in the PC computer was discussed, it is a sandwich of two metal plates on a quartz crystal of thickness t. The oscillation in practice is transverse, a shearing motion, the top layer moves to the right and the bottom layer moves to the left (Figure 3.1), and so the formula involves the shear modulus S. The formula we found is $f = (1/2\pi)(1/t)(2S/\rho)^{1/2}$, where t is the thickness of the layer. An approximate value is 0.27 GHz at 5-μm spacing in quartz. This oscillator frequency is then multiplied in circuitry to obtain the actual clock frequency perhaps 1.8 GHz. Beyond the limit of scaling we note that the frequency of oscillation of the hydrogen molecule is 1.3×10^{14} Hz, i.e. 130 THz.

We found in Figure 3.5 that vibrational frequencies scale as $1/L^n$ over 14 orders of magnitude, with $n \approx 1.32$! In the text Eq. (3.3) (see note N3) explains that $n = 2$ is expected for scaling in one dimension, while $n = 1$ is expected for isotropic scaling. In an overall summary, the oscillation frequencies of the oscillators described in the text behave in an intermediate fashion.

In the realm of motion of objects in viscous media like air a transition in behavior occurs around 1-mm size, such that an airplane will no longer glide but it will simply fall. A bumblebee does not "fly" but rather it pushes the air down with its wings, like oars of a boat in water.

We mentioned briefly that the rule of scaling for the FET devices is described as constant electric field scaling.

References

1 Nadar, K. (2011) Shear modulus, http://en.wikipedia.org/wiki/Shear_modulus (accessed 5 December 2011).

2 Nadar, K. (2011) Picture of quartz crystal (tuning fork type), http://en.wikipedia.org/wiki/File:Inside_QuartzCrystal-Tuningfork.jpg (accessed 5 December 2011).

3 Carr, D.W., et al. (1999) *Appl. Phys. Lett.*, **75**, 920.

4 Reuckes, T., et al. (2000) *Science*, **289**, 94.

5 Suri, C.R. (April 2006) Quartz crystal based microgravimetric immunobiosensors. *Sens. Transducers*, **66** (4), 543–552.

4
Biology as Successful Nanotechnology

Biology, the realm of whales and elephants and humans, may seem quite separate from anything involving nanoscale elements. On the other hand, biology also involves cells, and bacteria such as *E. coli*. We are also likely to think of genetics and DNA in connection with biology, which also makes contact with nanoscale phenomena.

The aspects we will discuss are motors, information technology, and sensors. Beyond this will talk briefly about self-assembly, replication, and evolution.

4.1
Molecular Motors in Large Animals: Linear Motors and Rotary Motors

Our discussion is going to be very simple. The first topic is motors. When we think of a large animal like an elephant or a whale we may wonder, where is the motor? There is no motor in the sense of having a V-8 engine or a nuclear reactor in a large animal. It may be surprising to realize that *all of the forces that propel these large animals come from billions of tiny molecules!* (Of course, the molecules are organized into regions, e.g., the bicep muscle in your arm.) There are two principal types of motors in biology, linear and rotary, and the kind of motor that we are talking about for large animals is a linear motor. Among linear motors, we will talk about myosin, which moves along actin. The muscle does its work by contracting and the contraction comes from the motion of myosin along an actin filament, which is about 5 nm in diameter, 1–4 μm in length.

Actin in the simplest terms is a bit like a nanoscale ladder. It is linear, it has rungs that are perhaps 5 nm apart and myosin, which moves along actin, is a bit like a monkey climbing a ladder. Myosin has two heads, and

---

Understanding the Nanotechnology Revolution, First Edition. Edward L. Wolf, Manasa Medikonda.
© 2012 Wiley-VCH Verlag GmbH & Co. KGaA. Published 2012 by Wiley-VCH Verlag GmbH & Co. KGaA.

it is known that the two heads on the myosin move "hand-over-hand," that the step length is about 5 nm, and that for each step an ATP molecule is taken in and an ADP molecule is released. An ATP molecule is about 2 nm by 0.7 nm, and the energy for the hand-over-hand walking motion of myosin comes from the chemical change of ATP to ADP. The "monkey" has a rope around his waist and he is pulling. On one end the rope is attached to a tendon, in a simplified view, and the other end is attached to a second tendon, and the resulting force applied to the tendon will pull, and, for example, rotate your forearm around your elbow joint as an axis. So if you lift your bicycle, for example, by using both arms and pulling upward, the bicep muscles in the upper arm contract and pull a tendon that causes the forearm to rotate and to lift the bike upward. This is simplified, but the active element is the molecular motion of the myosin along the actin, which is akin to a nanoscale monkey climbing a ladder. The myosin head, the "hand of the monkey," grasps one rung and then swings the other hand, the other side of the myosin dimer molecule, around to grasp the next rung. The steps and motion are 5–8 nm in length and the energy for this comes from changing chemical molecule ATP, which is "adenosine triphosphate," and as the motion occurs, the triphosphate is changed to a double phosphate ADP giving up energy the energy goes into pulling the rope along the actin ladder. It is reported that the speed is 1–4 μm/s, so the step rate is 120–800 Hz. These actin filaments are organized into bundles, for example, a striated muscle cell may be 10 μm in diameter, and it is complicated, but the fact remains that the only source of motion is molecular contraction.

The other principal form of *linear molecular motor* is kinesin moving along tubulin, similar to myosin moving on actin, in size scales, in mode of action, and in transformation of ATP to ADP.

To get *macroscopic* motion these contracting elements have to be added together in series. If we connect two in series, each contracting at 1 μm/s, then the outer ends would be contracting at 2 μm/s. So to get overall speed of muscle contraction say 0.1 m/s would require a large number, N of units in series. We can determine the value of N by setting

$$N \times 4 \times 10^{-6} \text{ m/s} = 0.1 \text{ m/s}$$

This implies $N = 25\,000$, the number of actin filaments in series. Since actin filaments are 4×10^{-6} m long, the overall length would need to be

$$25\,000 \times 4 \times 10^{-6} \text{ m} = 0.1 \text{ m}$$

The predicted length is 0.1 m (10 cm), which is reasonable, similar to the length of a bicep muscle.

These motors work by changing chemical energy into work, defined as force times distance. In conversion the efficiency is perhaps 50%, which is high, compared, for example, with the efficiency of the internal combustion engine, which might be 20%.

The second kind of motor in biology is the *rotary motor*. Rotary motors appeared very early in the development of organisms. Bacteria, such as *E. coli*, swim through water using propellers, called *flagella*, which are turned by rotary molecular motors. *E. coli* are single-cell organisms, about 3 or 4 μm long with a diameter of 0.6 μm. So *E. coli* resembles a 600-nm-diameter submarine. It is a self-contained system with an outer wall, called the lipid wall. Through the cell wall there are perhaps 10 rotary motors, which have rotary shafts oriented perpendicular to the wall, on the outside attached to propellers that are rotated and cause the bacterium to move in its environment, which is water. The speed of swimming of *E. coli* is perhaps 10–20 μm/s, and it thus moves two or three times its length per second. (A picture of a similar bacterium is shown in Figure 5.1.)

A rotary molecular motor is indicated [1] in Figure 4.1.

In Figure 4.1, the energy source is indicated as ATP, which converts to ADP.

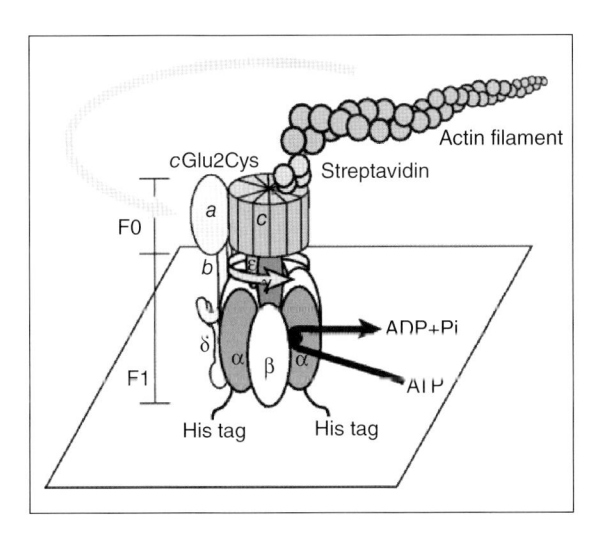

Figure 4.1 Fluorescently labeled actin filament permits observation of rotation of the c subunit of the molecular motor. The top portion ("streptavidin" and the actin filament) was added by researchers to observe the rotation. The γ, ε, and c units are thus shown to be a rotor, while the α, β, δ, a, and b complexes are the stator. The rotation rate of the actin filament in the viscous medium was found to depend upon its length. Rotational rates in the range of 0.5–10 Hz were measured, consistent with a torque τ of 40 pN nm (Sambongi et al., Science 286, 1722 [1999]).

The rotary molecular motor appears in various bacteria and is a unit that evolved early on and has been retained in biology to the present. The rotary motor (see Figure 4.1) is about 8 nm in diameter, and 14 nm long, sticking through the cell wall. The shaft is 0.8 nm in diameter and rotates at about 5 Hz! The motor also depends upon taking the energy-rich ATP molecule and changing that molecule to double phosphate ADP, and this produces rotation in this case. The torque produced by this rotary motor is in the range of 20 pN nm = 2.0×10^{-20} N m (1 pico = 10^{-12}).

These motors, which occur naturally in living cells, such as E. coli and magnetic bacteria that we will mention, can operate separately from the living cell. Experiments [2] performed at Cornell University involved extraction of the molecular motors from living organisms, and growing many more using nutrient solution in the lab. Rotary motors were thus removed from the living cells and a large number of the motor molecules were produced. The molecular motors were deposited on an array of Ni metal posts, 200 nm high and 80 nm in diameter, spaced by about 2.5 µm, prepared by nanotechnology, in an experimental situation. The motors were equipped on the top with specially fabricated "nanopropellers," 750–1400 nm long and 150 nm in diameter. (In the Cornell work, the metal plates replaced the "actin filaments," which are shown in Figure 4.1.) These propellers were put onto the motors in place of the flagellum (the natural propeller element that, when rotated in water, pushes the bacterium forward).

The propellers in the experimental setup were observed by an optical microscope of 60–100 magnification and the events were recorded with a video camera.

Rotation of some of the motors was observed, but only when the solution contained ATP, some of the motors would turn, they all turned in the same direction. If the ATP was removed from the solution, all rotation stopped. The frequency of the rotation, about 5 rps (rotations per second), could be observed by microscopic observation, and the amount of work done by the rotating nanopropellers moving through the water could be estimated based upon the propeller dimensions and the viscosity of the water. Such a motion, resembling that of an oar through a viscous fluid like water, is well understood, and can be modeled numerically. This observation and analysis confirmed that the efficiency of conversion of chemical energy ATP to ADP to work done by the rotation of the propeller in the viscous medium was about 50%. The torque of the motor was estimated from the observations and was about 20 pN nm. The authors of the study [2] suggest that at least some of the motors actually rotated as long as 2.5 hours, which would correspond, at rotation of 5 rps, to 45 000 rotations! The bearing on this molecular motor seems to be pretty strong! The workers estimate that energy release by ATP is about 80 pN nm per molecule. In eV (electronvolts) units this is 0.5 eV per molecule.

How big is the torque of the rotary motor, 20 pN nm? A physicist might remark that a torque produces a change of angular momentum, and a famous atom scale angular momentum is $\hbar = 1.0 \times 10^{-34}$ N m s ($\hbar$ is the reduced Planck constant), so this molecular motor would make a rate of change of momentum equal to $1.9 \times 10^{14} \hbar$/s! So on an atomic scale this molecular motor torque is huge, but on most practical scales it is very small. For a comparison, E.W. is a bike rider, and if he stands up, all his weight on one bike pedal, at radius 0.25 m, force = 150 lb = 667 N, then the torque = $F \times r$ = 167 N m. So this human-size torque is 8×10^{21} times bigger than the torque of the molecular rotary motor.

What is the power[1] of the rotary motor? Using the above numbers for torque and frequency of rotation and the formula: power $P = Fv = \omega\tau = $ (omega) × (torque), we find $P = 2\pi \times 5 \times 2 \times 10^{-20} = 6.28 \times 10^{-19}$ W. This is tiny, but suppose we can assemble many of them. We can think of powering a Honda Civic, at 100 HP, with N of these motors. Since 746 W = 1 HP, we get N= $100 \times 746/6.28 \times 10^{-19}$ =1.19×10^{23} molecular motors.

This is quite analogous to what happens in a large animal, like an elephant, but those motors are primarily linear motors. As we have said, the only source of motive force in large mammals is the concerted action of large numbers of molecular units, such as myosin moving on actin filaments. To get 100 HP, we find that 1.19×10^{23} of the rotary motors will be needed. Overlooking how we will connect the motors, we can estimate the total volume occupied. The volume occupied by the molecular motor, from the rough dimensions given above, is about 896×10^{-27} m³, so the required volume of the scaled-up Honda Civic "molecular motor" would be 0.106 m³, for example, a cube of side L = 47.3 cm.

So this is pretty promising! The weight of the 100 HP motor would be quite small, because the density of the bacteria is nearly the same as the density of water, since it floats in water. On this basis 1 g cm^{-3} = 1000 kg/m³ we get mass 106 kG = 2.205×106 = 233 lb. So to fill the gas tank we need to buy some ATP!

These motors are molecules, they do not have to be alive, and they can be extracted from living matter. If they could be produced in large numbers to be placed in a nanostructure they could lead to motion of larger objects, and as we have said, the cooperative motion of perhaps 10^{22} (say 10 HP, 7460 W) of such molecular motors (in their linear form) causes the motion of large animals like whales and elephants.

4.2
Information Technology in Biology Based on DNA

Another aspect of nanotechnology that is superbly accomplished in biology is the control of information. Information in biology is encoded in DNA. For our purpose DNA, which is known since 1953 from the famous work of James Watson and Francis Crick in Cambridge, UK, to be a double-helix structure or double-strand structure. (Sketches are shown in Figures 4.2 and 4.3.) For our purpose we can think of a single strand of DNA, it is again a ladder, now a flexible rope ladder. So it has rungs, like the actin, but is flexible, and the rungs on the rope ladder, which we think of as a single strand of DNA, the rungs on this rope ladder are spaced by about 0.5 nm.[2] Each rung on the rope ladder that we associate with a single strand of DNA has a chemical molecule called a base. This is a molecule and there are four different choices for the base molecule. These are similar molecules, about 0.6 nm in diameter, and they are called A, G, T, and C. Along a single ladder, we can have at each rung any one of these four molecules, A, G, T, or C. *But the base molecules have interlocking properties.* A and G join together and T and C join together. In a very specific way, so A and G are like a lock and

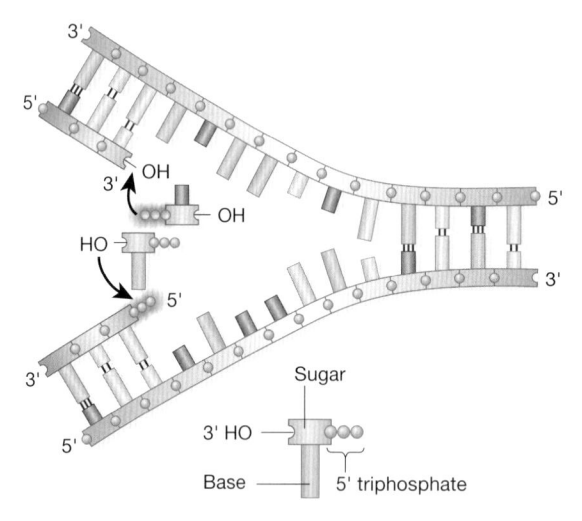

Figure 4.2 Models of DNA replication fork, showing breaking of hydrogen bonds between upper and lower bases (center of figure) as DNAP engine (left) pulls double-helical DNA from right to left. Hydrogen bonds are depicted here as smaller-diameter cylinders connecting the larger cylinders representing the four different bases A, C, G, and T, which, however, bond only as complementary pairs AT and CG. The AT pairs form double hydrogen bonds, while the CG pairs form triple hydrogen bonds (B. Alberts, Nature 421, 431 [2003]).

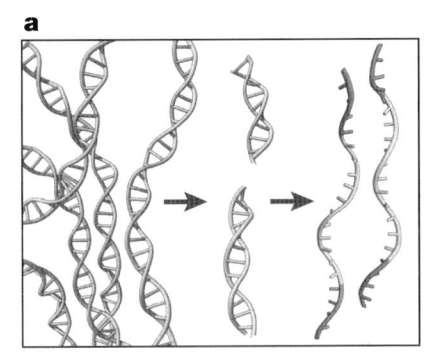

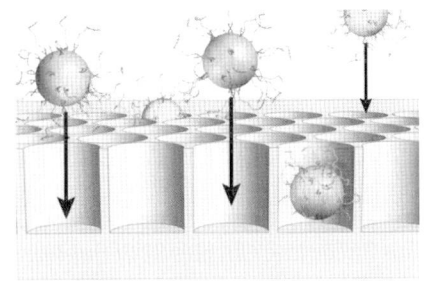

Figure 4.3 (Left) Genomic DNA is isolated, fragmented and separated into single strands. Not shown, single-strand fragments are bound to 28-μm-diameter beads under conditions that favor one fragment per bead. Next, cloning of the fragment on each bead multiplies the identical fragment so that 10^7 identical fragments attach to each bead. (Right) Beads, each holding 10^7 identical fragments (but base ATGC sequence differs from one bead to another) are loaded into fiber-optic wells ("picoliter reactor cells"), in an array of 1.6 million wells, each at the head of an optical fiber. Beads are sized 28 μm such that one bead fills a well. These are imaged by a 16-megapixel CCD camera (each well is imaged by about 9 pixels). Not shown, A,T,G,C reactants sequentially flood each well, releasing 10 000 photons per base per well. This record provides the sequence on each bead. Nucleotides (bases) are detected by the associated release of inorganic pyrophosphate and the generation of photons (Margulies et al. Nature 437, 376 [2005], Figure 1a).

a key, and T and C are also like a lock and a key, but neither A nor G will stick to T or C. So the only pairings possible are between A and G and between T and C. So we can imagine that if an initial strand, an initial rope ladder with whatever sequence of bases along its rungs, if we allow that rope to float in a solution with all four bases present, we will find each base in the original rope will pick its complementary base out of the solution. This will allow then construction of a double ladder, with the A and G pairing and the T and C pairing actually binding the two rope ladders together. A given sequence on a single strand can then easily generate its complementary single strand, and the result is double-strand DNA.[3] The double helix is the joining of two rope ladders and it occurs only when the first ladder finds on the second ladder the complementary base at each rung. If we sequence AGTC on one side, opposite side must be GACT, and this will stick together. The complementary base for A is G, etc., lock and key. How does this store information?

When we have the double helix, the two ropes bonded together at each rung on the double rope we will have either AG or TC on successive rungs. This is like binary information. AG is spin up and CT is spin down, or AG is one and CT is zero. These are the "bits" in the information sense, of the biological information system. Each rung is *one bit* of information. *This is a binary data-storage system.* We can investigate the information density in this double-ladder scheme.

The spacing between the rungs, let's assume, is 0.5 nm. If we assume the diameter is 2 nm, then we can find the density of bits per unit length, evidently about 2 Gbits/m. A TB is 10^{12} bytes = 8×10^{12} bits. At 0.5-nm spacing *1 TB is stored in 4 km of double-strand DNA.*

DNA is flexible, and likes to curl up into a spiral or ball up into a spherical shape, because of van der Waals attraction forces (see Chapter 9). If the 4 km length, representing 1 TB, assuming the diameter of the DNA molecule itself is 2 nm, curls into a planar spiral, like a pancake of radius r (and thickness 2 nm) we find[4] $r = 1.6$ mm, which is much smaller that a TB disk drive, which is several inches in diameter. The areal density is $1\,TB/(\pi r^2) = 80.2\,TB/(in)^2$. This is eight times the Library of Congress in 1 square inch.

If we allow the 4 km length DNA, encoding 1 TB, to ball up into a sphere[5] of radius R (so $V = (4/3) \pi R^3$) then $R = 1.44 \times 10^{-5} = 14.4\,\mu m$. This represents extremely dense data storage, about $7.94 \times 10^{13}\,TB/m^3$. If the nucleus of a mammalian cell has diameter 6 μm, then on this basis it could contain up to 8.97 GB of information, if fully occupied by DNA.

A single strand of the CGAT "rope ladder" can be induced to produce a complementary strand, and the two will form a stable double strand. The strands can be peeled apart, and having one available will allow a complementary one to be constructed.

The cell has inside it means for replicating its DNA. In such a process, the DNA of the system is uncoiled, reduced to single-strand form, which with its exposed sequence of bases, can generate the complementary strand. This can be sent out, to represent a copy of the information of the system. The information, once established in the CGAT order along a given strand, can be copied and sent elsewhere. This information transfer is the magic of biology as a nanotechnology. Such capabilities are not available in man-made nanotechnologies.

The unit in the cell that copies DNA is called the "DNA replication fork" and a sketch of this is shown in Figure 4.2.

Let us talk how about the information is embedded in the cell, we will talk about E. coli, the best-studied bacterium.

In the structure of the cell, we have mentioned the outer wall, through which the rotary motors are mounted. We now focus on a second interior part, called the nucleus, which may be 6 μm in diameter. The nucleus in the cell contains its information, so that the DNA that represents the cell is located in the nucleus. We have learned DNA is long, but can coil up to form spirals and balls, and the DNA in the nucleus of the cell will be in a balled-up form to save space. The nucleus facilitates the replication of the cell DNA, and then the cell itself. The magic feature of biology is, not only is it self-assembling, but it is self-replicating. The replicating starts with one E. coli and ending up with two E. coli bacteria (see Figure 14.2 for replication of an artificial bacterium). This is accomplished in large part within the nucleus of the original cell. The DNA strand that describes the cell is copied. It appears that the DNA required for E. coli is on the millimeter scale in length, containing 4.72 million base pairs and that the copying process in the cell nucleus takes [3] about 50 minutes! The copy of the DNA then facilitates a copy of the whole system, which is the whole bacterium. Although the bacterium is small, it is complex in its structure and operation.[6]

Finally, in connection with DNA, what are chances for change in the DNA, leading to changes in the cell or larger organism? The DNA of the system is the blueprint for building that system. We have heard of clones. Well, a clone is an organism constructed from the DNA of the original organism. The DNA contains all the information about the cell, or about the larger organism, which could be a dinosaur or a sheep! Changes in the DNA may result from damage of various kinds. The binding of the AG and CT is not strong on a chemical bonding scale. A bond in silicon is ~1.4 eV, the bonding AG and CT that preserves the information in the biological cell is much weaker. The importance is that bonding be strong enough to exceed $k_B T$ the thermal agitation energy. Since $k_B T$ for living organisms is not much higher than 300 K, with $k_B T$ about 26 meV, energies less than 1 eV are good enough to keep the AG and CT bonds intact. Still, mutations, changes in the DNA, do occur. The famous understanding obtained by Charles Darwin was that accidental changes in DNA, mutations, may produce a better organism with respect to its environment, and that organism may become dominant. This is the basis for evolution, the change in the properties of organisms and species.

An important frontier in biology at the present time is "DNA sequencing," which means to detect and write down (decode) the information in the DNA of a given organism. DNA sequencing is basically biology, but according to the Royal Society Report [4] it can be considered "an emerging nanotechnology," because it can be done more rapidly using nanotechnology. We saw in Figure 1.1, a map of human migration, based on DNA sequencing. Human DNA samples were obtained from people all over the world, and these samples were "sequenced" with particular emphasis on the "Y-chromosome" and "mitochondrial DNA." These parts of the human genome are useful to track migrations of *Homo sapiens*, because these portions are passed from parent to child identically, unless there is a mutation. The Y-chromosome is present in males and is passed from father to son. If there is a change in the Y-chromosome it represents a mutation, and the changed DNA is thereafter retained. So if change A in the Y-chromosome first appears in Siberia, and the same specific change A is found in *Homo sapiens* in Northern Europe, perhaps with an added change B, the inference may be possible that the Northern European humans had migrated from Siberia. (A sequence ABCD of DNA changes can provide a record of migrations, and extrapolation back from A to no mutations can suggest where the species originated.) A second useful part of DNA is associated with mitochondria, related to energy production in cells. The mitochondrial DNA is passed identically from mother to child, male or female, and again changes in this portion of the human genome can be attributed to mutations, which will persist in later generations. An approximate method of dating the mutations, based on an average rate of mutating, called "the molecular clock" was first described by Rebecca Cann *et al.* [5]. It is found that going back to earlier and earlier generations (fewer mitochondrial DNA mutations) that the common maternal ancestor was about 200 000 years ago in Northern Africa, this is said to be "the mitochondrial eve" of humanity.

Sequencing of DNA is difficult and has evolved steadily from its origins as brilliant and difficult biochemistry, most recently by incorporating methods of microelectronics [6]. In its present form DNA sequencing has been described, as we mentioned, as an "emerging nanotechnology" [4]. A recent method of sequencing is described [6] in Figure 4.3.

In the work of Margulies *et al.* [6] the basic approach is to break down the DNA into fragments of several hundred base pairs, to isolate individual fragments onto carrier beads, and to amplify by DNA replication to about 10 million identical fragments per bead. Individual beads are loaded into micrometer-sized wells ("picoliter reactors") where they can be exposed in sequence to reactants containing the bases ATGC. Brilliant chemistry allows sequencing of the fragments by observations of photon emissions (fluores-

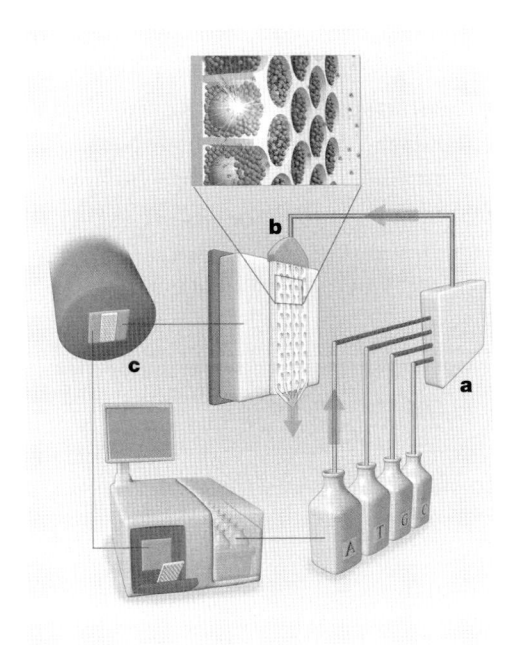

Figure 4.4 Sequencing instrument, consists of (a) fluidic assembly providing sequential flows of reactants (b), flow chamber includes the fiber-optic slide with 1.6 million individually imaged picoliter reaction chambers (top, showing flow of reactants past the wells); (c) CCD imaging system and (d) computer system which operates in real time (Margulies et al. Nature 437, 376 [2005], Figure 2).

cence), specific to each fragment's sequence and to the reactant choice, which are recorded by a CCD camera.

The authors [6] indicated that their adoption of modern methods puts the DNA sequencing onto a trajectory similar to that of the Moore's law advance in computer chips, in which succeeding generations of devices have larger numbers of cells, greater processing speed, and smaller cost per cell. It is expected that the cost of sequencing the human genome will be substantially reduced as methods such as pioneered by Margulies et al. are refined. The role of nanotechnology here [1] is to improve the performance of methods that originate in biochemistry. The methods are also dependent on computer power and elegant programming (Figure 4.4).

In sequencing a 580 000 base genome, the Margulies et al. [6] apparatus provided 42 complete cycles of the four nucleotide reagents (bottom right, figure) in 243 minutes. The average fragment length was 108 bases.

In biology, the "information technology" in DNA has been essential to self-replication and gradual improvement by evolution, and this aspect is thus-far absent in man-made nanotechnology. Man-made nanotechnology can produce dense data storage, however, far short of the data density of curled up DNA. Man-made nanotechnology can in fact do things that biology certainly cannot do. In biology nothing happens faster than a millisecond (corresponding to a frequency of 1000 Hz), whereas man-made computers commonly work at 3.0 GHz. So the speed of operations in electronic nanotechnology is orders of magnitude faster than anything that occurs in nature. But man-made nanotechnology, it seems, has no hope of being self-replicating in the useful fashion that occurs in biology.

4.3
Sensors, Rods, Cones, and Nanoscale Magnets

Sensors are another type of nanoscale device well developed in biology. Sensor cells include rods and cones in the eye that detect light, and cells in the ear that detect sound vibrations.

Another sensor not so obvious in animals but is probably present is the magnetic field sensor. Magnetic field sensors are known in bacteria. The magnetic bacteria is 3 or 4 µm long, it lives at the bottom of the ocean where there is no light, and the magnetic field sensor aligns the bacterium along the magnetic field of the earth that is approximately vertical as it comes out of the ocean floor. Transmission electron microscope studies of magnetic bacteria show that they have nanometer-scale magnets along the axis of the bacteria. An example showed an array of twenty-two 40-nm-size magnetite Fe_3O_4 magnets spaced closely together in a linear array. Altogether this constituted a bar magnet strong enough to align the bacterium along the magnetic field of the earth, 1 or 2 Oersteds or 10^{-4} T. It is clear that the bacteria determine the correct direction to swim, that is, down to the bottom of the ocean, by the orientation that the magnet provides. If the bacteria that are prevalent in the southern hemisphere are moved to the northern hemisphere, where the magnetic field direction is reversed, these bacteria are observed to swim in the wrong direction. They depend upon the sensor to align them along the field and the organism has learned which way to go along the magnetic field, depending on where it lives. These bacteria, put into a solution can be seen to swim in one direction and to turn around if the local magnetic field is reversed. This is a sensor, a nanometer-scale object, and it is an illustration of growth of ferromagnetic crystals from solution. The growth must occur from iron obtained in the ocean where the bacterium grows, and the growth of these crystals, magnetite and a similar sulfur compound, occurs completely in an aqueous environment. No semiconductor technology is involved!

4.4
Ion Channels: Nanotransistors of Biology

Ion channels are molecular devices that we can think of as the nanotransistors of biology. These are devices in the walls of cells that monitor and control the passage of nutrients from the outside to the inside of the cell.

The smallest forms of life are bacteria, which are single cells of micrometer size, which are enclosed by an impermeable lipid bilayer membrane, the cell wall. This hydrophobic layer is akin to a soap bubble. Lipid cell walls are ubiquitous in all forms of life. Communication from the cell to the extracellular environment is accomplished in part by ion channels, which allow specific ion species to enter or leave the cell.

Two specific types of "transmembrane protein ion channels" are the Ca^{++}-gated potassium channel, and the voltage-gated potassium-ion channel, which is essential to the generation of nerve impulses. The dimensions of these transmembrane proteins are on the same order as their close relatives, the rotary engines, which were characterized as 8 nm in diameter and 14 nm in length.

These ion-channel structures are "highly conserved," meaning that the essential units that appeared about 1 billion years ago in single cells, have been modified, but not essentially changed, in the many different cellular applications that have since evolved.

In this chapter we gave examples from biology of nanotechnology in the form of motors, information storage and information handling, and a bit about sensors and ion channels, the nanotransistors of biology.

References

1 Sambongi, Y., et al. (1999) *Science*, **286**, 1722.
2 Soong, R.K., et al. (2000) *Science*, **290**, 1555.
3 Goodsell, D.S. (1993) *The Machinery of Life*, Springer Verlag, New York, p. 63.
4 The Royal Academy of Engineering, The Royal Society of London (2004) Nanoscience and nanotechnologies: opportunities and uncertainties. Royal Society Report 2004, http://www.nanotec.org.uk/finalReport.htm (accessed 3 December 2011). (See p. 20, array technologies.)
5 Cann, R.L., Stoneking, M., and Wilson, A.C. (1987) Mitochondrial DNA and human evolution. *Nature*, **325**, 31.
6 Margulies, M., et al. (2005) *Nature*, **437**, 376.

5
The End of Scaling: The Lumpiness of All Matter in the Universe

We have seen that a central theme of nanotechnology is to reduce the size scale of devices. Often, as in the Pentium chip, we get more devices that work as well or better than larger devices.

The question is, how far can such a useful scaling or miniaturization process go? What is the limit of scaling? The answer is that the essential lumpiness of matter eventually limits this process, in the 1–100 nm range, but at the same time brings into play new principles that can lead to new devices.

5.1
Lumpiness of Macroscopic Matter below the 10-μm Scale

We have the idea from looking at a glass of water that matter is smooth and continuous. We get the same impression by thinking of the glass itself; it looks perfectly smooth, structureless, with no particles anywhere to be seen. However, if we look at smaller and smaller size scales, we will find that the material is lumpy.

Let's do a thought experiment. We have heard of Captain Nemo, who was the captain of the Nautilus, a large submarine, in the Jules Verne book, "20 000 Leagues under the Sea." Let's scale Captain Nemo. Of course we can't scale a human being, but we can think of scaling his submarine. His submarine is probably not so different in size from a modern nuclear submarine, which we might say is 100 m long, completely self-powered. Suppose we imagine a robot submarine and scale it from 100 m down to a few meters. That would lead us to the robot submarines that are used in the Gulf of Mexico to work on undersea oil well-heads. Nothing will change in going from 100 to 10 m. In fact, if we smoothly scale our robot down as far as a micrometer, 10^{-6} m, the submarine should still work. An example of such a

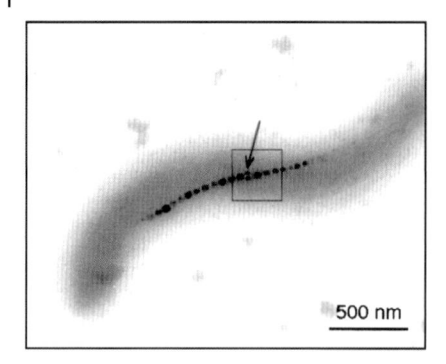

Figure 5.1 A single bacterium of length a few micrometers, able to swim with propellers (flagella, not visible) that are turned by molecular motors. This working "submarine" is in the size scale where Brownian motion is probably noticeable, an indication of the lumpiness of water. This is an image of magnetotactic bacteria, showing a linear array of magnetic crystals (Dunin-Borkowski et al., Science 282, 1868 [1998]).

submarine is a bacterium. We have seen in Chapter 4 that bacteria are a few micrometers long; they are self-contained, moved by propellers turned by molecular motors. Instead of nuclear power these bacteria have a metabolism, they eat, they create energy-rich molecules called ATP (adenosine triphosphate) which then turn the motors of their propellers. Some bacteria have sensors. (See Figure 5.1 [1].) The bacterium called *M. agnetospirillum magnetotacticum* grows internally an array of nanometer-scale magnets, made of the magnetic minerals Fe_3S_4 *greigite*, and Fe_3O_4, *magnetite*. These 40-nm-sized magnets, 22 of them in an array 1.2 µm long, align the bacterium along the magnetic field of the earth, pointing it in the right direction to swim to get to the bottom of the ocean, where the food is located. There is no light at the bottom of the ocean, nor any sense of direction from gravity, so these magnetic sensors are essential to the survival of the bacteria. A magnetic bacterium viewed as a controllable robot could easily be steered by providing a small mT magnetic field! So going to 3 µm (0.3% of a millimeter) does not rule out a functioning "robot submarine" moving through the water. Magnetite compasses were invented in China about 200 BC and were widely used on Chinese ships by 1100 AD.

Figure 5.1 is a typical bright-field transmission electron microscope (TEM) observation [1] of a single cell of the bacteria *M. magnetotacticum*. This image [1] emphasizes the "magnetite chain 1200 nm long containing 22 crystals that have average length and separation of about 45 and 9.5 nm, respectively." Many bacteria, such as *E. coli*, look much like this, apart from the magnets.[1]

So going to a micrometer scale will not prohibit the motion of the imagined robotic submarine in moving through the water, although an external observer of its motion might get seasick! (The external operator of the submarine, a bit like the operator of a drone aircraft, with a screen in front of him, would be insulated from the buffeting but might see erratic motions.) For even in the range 20 µm and below, the robot submarine will be subject to inherent turbulence, known as Brownian motion. The buffeting that will occur is perhaps not so different from the turbulence that is familiar in a modern jet airplane. The captain will say, fasten your seatbelts, and there will be some jolts. The jolts that occur are from collective motions of the air that bump the airplane around. The scaled submarine at sizes below 20 µm will experience a similar kind of jolting, called *Brownian motion*, the first sign of the lumpiness of water, in this case. Jolting motion of suspended pollen grains in water was first observed by the famous Scottish botanist Robert Brown in 1827, using a microscope such as was earlier employed by Anton Von Leeuwenhouk in 1674, a one-lens device through which he first observed bacteria. Brownian motion was interpreted in detail in terms of thermal equilibrium properties of a fluid by Albert Einstein in 1905 as jolting of a suspended object by impacts from collective motions of molecules in the supporting fluid.

If we now continue our imagined scaling of Captain Nemo's submarine the Nautilus below 1 µm, going down to 100 and 10 nm, certainly there will be huge changes. The imagined robot submarine itself will be lumpy in its construction, the situation is not so different from the end of Moore's law, with a final conclusion that you cannot make a submarine nor a transistor out of a single atom. So Brownian motion is the first indication of the essential lumpiness of matter.

5.2
Hydrogen Atom of Bohr: A New Size Scale, Planck's Constant

Let's now take a fresh look at what has been learned about matter at the nanometer scale and below. The Universe, as scientists now believe, is made up of particles and detailed mathematical rules that govern the motion of those particles. These particles are photons, electrons, protons, and neutrons, and we will now talk about how they combine to form the essential lumps of matter that are atoms.

We start with photons. Perhaps surprisingly, after the discovery of the electron in 1897, light was the first phenomenon to be discovered as having an essential lumpiness to it. The famous physicist Max Planck in 1900 concluded that light had to come in tiny chunks, and the size of the chunks, measured in energy is $E = h\nu$. The Greek letter nu, in the physics literature is often the symbol for frequency (also represented as f), so it is measured in cycles per second, or hertz. The constant h, which was measured by Planck, is extremely small, with the value $h = 6.6 \times 10^{-34}$ J s. Planck did not invent or derive this number, but found it by fitting theory to data. This number, $h = 6.6 \times 10^{-34}$, is built into the Universe that we live in. Planck came to this surprising conclusion by careful analysis of the spectrum of light emitted by hot objects, for example, the sun. Sunlight is composed of different colors, corresponding to different frequencies v, as was known to Newton who used a prism to disperse the light. Planck was able to quantitatively, numerically, model the sun's light spectrum, that is, how much energy was available at each of the frequencies, assuming energy was hv, as we have described. Planck found that his formula would also predict radiation spectra as a function of temperature, to include, for example, the gentle red glow you see looking into a pizza oven. In this formula, again, v is the frequency of the light, and is related to the wavelength λ as $v = c/\lambda$, where c is the speed of light, 3×10^8 m/s. Planck's fundamental discovery was that light comes in quanta, called photons, whose energy is hv. The number for h was obtained by fitting to the data, so that $h = 6.6 \times 10^{-34}$ J s is an experimental quantity.

The fundamental nature of Planck's discovery in 1900 was confirmed in 1905 by the discovery of the photoelectric effect. When light falls on a substance, such as a piece of metal, electrons can be emitted, and in fact what happens is that a single photon will disappear to release an electron at a total energy cost that exactly matches the energy $E = hv$, with $h = 6.6 \times 10^{-34}$ J s, lost from the photon. This entirely different experiment gave *the same value* of Planck's constant.

It is important to understand the role of Planck's constant in nature. It is an observed quantity, measured by fitting of black-body spectra and also by the photoelectric effect observations. Its value cannot be predicted by any kind of theory. However, as we will see, one might suggest that Planck's constant instead of being called h, might be better called A, for Atlas! Atlas is the Greek god who holds up the world. In a moment we will see in what sense Planck's constant can be said to "hold up the world."

The electron was discovered by J.J. Thomson, at Cambridge University in England. He measured the mass of the electron, which is 9.1×10^{-31} kg, by observing deflection of moving electrons in a magnetic field. The charge of the electron was measured as 1.6×10^{-19} C by the American physicist Millikan in 1911. The proton, discovered by Ernest Rutherford in 1918 following his discovery of the atomic nucleus in 1911, is 1836 times heavier than an electron, has a positive e charge, and is now known to have a radius of ~0.8 fm. A femtometer, 10^{-15} m, is a million times smaller than a nanometer.

It was discovered a bit later that the electron and also the proton each have a *magnetic moment* μ.[2] Electrons and protons act like small bar magnets, and these particular magnets can have only two orientations, *up or down* in a magnetic field, a property of spin ½. (We will learn later more details about the states of a spin.) The electron of charge e also has an inherent spin angular momentum, which is precisely $h/4\pi = \hbar/2$. The basic relation is that magnetic moment is $\mu = (e/2m) L$, where L is angular momentum, e is electron charge, and m is mass.

How do these simple particles form atoms? Hydrogen, the simplest atom, consists of an electron and a proton. Planck's constant h played a central role in a model for the hydrogen atom. Neils Bohr, the famous Danish physicist, in a 1912 letter to Ernest Rutherford, explained a wonderfully useful simple model of the hydrogen atom. In this model, *the electron orbits around the proton*. One of the famous rules of nature is that opposite electric charges attract. The energy of attraction of the electron and the proton is $U = -k_c e^2/r$, where k_c, the Coulomb constant is 9×10^9 N m^2/C^2, e is the electron charge, and r is the spacing between the two charges. Bohr found that if he made the angular momentum of the electron around the proton equal to $n\hbar$, with n an integer, then he could explain many features of the hydrogen atom, including its observed sharp absorption and emission wavelengths for light. This was a remarkable feat. He found, by making this bold assumption, the exact formula for the inherent radius of the hydrogen atom,

$$a_o = \hbar^2/k_c me^2 = 0.0529 \text{ nm} \qquad (5.1)$$

Bohr's radius, the atomic size, is *proportional to the square of Planck's constant* $h/2\pi = \hbar$. If Planck's constant were to go to zero, then the hydrogen atom and all other atoms would collapse! Planck's constant is needed in our Universe to hold up the atoms[3] that hold up our world. This is a bit like the role of the Greek god Atlas! Bohr found that the binding energy of the electron is 13.6 eV − $ke^2/2a_o$, which can be regarded as establishing an energy scale for atoms.

Bohr's radius $a_o = \hbar^2/ k_c me^2 = 0.0529$ nm defines a new length scale, the atomic length scale. About 10 hydrogen atom diameters make up 1 nm. So the end of scaling is the atomic length scale, the Bohr radius, about 1/20 of a nanometer.

The hydrogen atom also produces new timescales, defined from the orbital motion of the electron, the atom is the basis for "atomic units," which are used by physicists.[4]

Now let's review the importance of Planck's constant. In Bohr's model, which is verified in its basic results by all refined theories, the radius of hydrogen is proportional to the square of h, Planck's constant. So if Planck's constant were decreased to zero, the radius of the hydrogen atom would go to zero. Planck's constant is small, 6.6×10^{-34} J s, but it is big enough to keep hydrogen atoms and all other atoms from collapsing. So in this sense it might better have been called A, the Atlas constant, which holds up the earth. And by understanding nanophysics, the laws of physics that govern at scales below 1 nm, including electron and proton spin interactions in atoms, we have reached the starting point of modern timekeeping technology of atomic clocks.

5.3
Waves of Water, Light, Electron, and Their Diffractions

Let's discuss a special feature of the nanophysical world, which is the relation between particles and waves. We have said that the particle of light, the photon, has energy hv and can be emitted or absorbed in physical processes. How can we reconcile this with our background knowledge that light is a wave?

Wave phenomena are a bit confusing, so let's think of a simple example! Suppose a pebble is dropped into a lake. The result will be circular water waves moving out from $r = 0$, where at $t = 0$, the pebble was dropped. Suppose there is a leaf at radius $R = 1$ m, which will be bobbing up and down as the waves go past. How can we describe the motion of the leaf? Suppose the vertical displacement of the water is $y = (1\,\text{cm})\sin(kr - \omega t)$. The displacement y, of the leaf is $y = 1\,\text{cm}\,\sin(k - \omega t)$, since $R = 1$ m. (The sin$x$ function oscillates with period 2π in x, has amplitude 1, and starts at zero when x is zero.) If we wait a time τ, the period, the repeat time for the motion, then the argument $(kr-\omega t)$ must have increased by 2π. So $\omega\tau = 2\pi$ and $\omega = 2\pi/\tau = 2\pi v$.

Now let's imagine a snapshot picture (like Figure 5.2) of the wave pattern at time 10τ. The picture of the ripples is $y = 1\,\text{cm}\,\sin(kr - \omega\,10\,\tau)$, which is the same as $y = 1\,\text{cm}\,\sin kr$. (This is because $\omega\,10\,\tau$ is $10 \times 2\pi$, and this makes no difference in the value of the sin function.) The ripples are spaced by Δr, which must be the wavelength λ.

So the formula $y = 1\,cm\,\sin(kr - \omega t)$ describes a moving wave of wavelength $\lambda = 2\pi/k$ and frequency $v = \omega/2\pi$.

Now the speed of the wave is the speed of some fixed point on the wave, like the point where the displacement is zero, where $kr = \omega t$. The point r then is given as $r = \omega t/k$. So the rate of change of r, dr/dt, is just ω/k, which is the phase velocity, the speed of the wave.

Figure 5.2 Picture of wave with bobbing object (http://www.johnthehandyman.co.uk/water%20ripple.jpg).

Now that we understand waves we can better understand the relation between light as a *wave* and light as a *particle*.

A light wave can be represented by $E = E_o \sin(kx - \omega t)$, where E is the electric field vector, sin is an oscillating function of period 2π and amplitude 1, $k = 2\pi/\lambda$ and $\omega = 2\pi\nu$.

The speed of light is $c = 3 \times 10^8 \, \text{m/s} = (\varepsilon_o \mu_o)^{-1/2}$, where ε_o and μ_o are fundamental constants called, respectively, the permittivity and permeability of free space. The relation between the light photons and the wave is that E^2, *the square of the electric field in the light wave, predicts where the photons will be found.*

It is a familiar experiment in high-school physics to show light diffraction, a wave property, by shining a laser beam, for example, from a red helium-neon laser, through two closely spaced slits. The result is a diffraction pattern of oscillating intensity on the screen behind the two slits. The peaks of this diffraction pattern are the points where a photomultiplier tube counting photons will show the largest count rate. The photon counter will show *no counts* at the nodes of the diffraction pattern. This kind of measurement can be improved with a "channel plate," which is an array of tiny photomultiplier elements. It is experimentally clear that the electric field acts like a "wavefunction" for photons, it predicts where the photons will be found.

If we imagine the standing wave of light as will occur at the surface of a mirror, the electric field will be zero right at the surface of the mirror and will have periodic peaks and minima spaced at half-wave distances from the mirror. The peaks of the standing wave pattern are the places where the density of photons is highest and the minima are places where one will not find any photons. This situation, which we might call wave–particle duality, also extends to matter.

5.4
DeBroglie Matter Wavelength

We have seen that light, thought of as a wave, is actually made up of particles. A second aspect of this duality, is that *matter particles have a wavelength, called the deBroglie wavelength*. This wavelength is $\lambda = h/p$, where Planck's constant again appears and $p = mv$ is the momentum of the particle. It is an experimental fact that beams of electrons can be engineered by their wave properties. Electrons can be focused by magnetic lenses, as in a TEM, using the wave property, specifically $\lambda = h/p$. This is another example of the pervasive role of Planck's constant h in our Universe.

De Broglie's wavelength, $\lambda = h/p$, where p is the momentum is the product of mass times velocity, allows another way of looking at Bohr's assumption that the angular momentum L of the electron around the proton in its orbit is $n\hbar$ also means an integer number of wavelengths around the circumference. That is, $n\lambda = 2\pi r$ (since $nh/p = 2\pi r$ implies $n\hbar = pr = L$).

We mentioned the high-school physics experiment to show light diffraction by shining a laser beam, through two closely spaced slits. *An electron beam behaves in the same way!* A diffraction pattern appears, first seen by Davisson and Germer at Bell Telephone Laboratories, who explained the diffraction angles using $\lambda = h/p$. A sketch of electron diffraction is shown in Figure 5.3. (Electron diffraction is a common research tool in surface

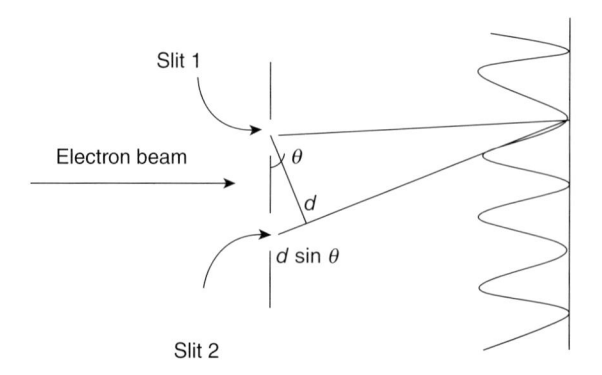

Figure 5.3 Sketch of electron diffraction in a two-slit geometry. The deBroglie condition $\lambda = h/p$ is found to predict the angles of maxima.

analysis, closely analogous to X-ray crystallography.) Here, of course, λ has units of meters, h has units of J s, and p has units of kg m/s.

5.5
Schrodinger's Equation

So matter has a wavelength $\lambda = h/p$. What quantity propagates in the matter wave? It is the wavefunction $\psi(x,t)$. Physicists use the Greek letter $\psi(x,t)$ as the *wavefunction* such that the square, $\psi^2(x,t)$, is the chance of finding the particle at location x and time t. The probability density $P(x,t) = \psi^2(x,t)$. This probability function has a normalization constant so that the integral (or sum of values over all possible locations) adds up to one.

If we want to predict the locations of particles like electrons, we need a scheme for finding the wavefunction for a particle in a given situation. The equation that determines the wavefunction ψ was discovered by Schrodinger, a German physicist, in 1926. *Schrodinger's equation* is based on conservation of the total energy of the particle. The total energy E has a kinetic energy term $\frac{1}{2} mv^2 = p^2/2m$, and a potential energy part U that often will depend on the local electric potential or voltage. In the case of the H atom, the potential energy of the electron near the nucleus is $U = -k_c e^2/r$. Schrodinger found that the momentum of the particle, $p = mv$, had to be related to the spatial rate of change of the wavefunction, $p = -i\hbar \, d\psi/dx$, where $i^2 = -1$, again uses Planck's constant.

We are going to stop here to offer a summary, and then a discussion of practical results of the topics in present technology. (We will return to learn more about Schrodinger's equation in Chapter 6.)

5.6
The End of Scaling, the Substructure of the Universe

At the end of scaling we find the fundamental lumps of matter: photons, particles, and their spins. These are governed by the new laws of nature typified by Planck's constant h, which keeps atoms from collapse.

We have learned in this Chapter 5 about Brownian motion, the first evidence of the granularity of matter. We learned that light comes in particles called photons, whose energy is $h\nu$, where $h = 6.6 \times 10^{-34}$ J s, and ν (or f) is the frequency in hertz. The value of h was found by Planck by fitting a formula to the observed spectrum of light emitted by a hot surface. We learned that the photoelectric effect experiment gives the same value for h, confirming that light comes in particles with a specific energy. Electrons and protons carry an intrinsic angular momentum, called spin, which comes in units $h/2\pi = \hbar$ called "h-bar." We learned about the Bohr model of the hydrogen atom, assuming the angular momentum of the electron around the proton is an integer multiple of $\hbar$. This model gives an exact simple formula for the Bohr radius, $a_o = 0.0529$ nm and established an atomic scale of length. The formula for Bohr radius has in it the square of $\hbar$, so that if $\hbar$ were reduced, atoms would get smaller. The basic energy of the electron in the hydrogen atom is $-E_o/n^2$, where n is an integer. The basic energy $E_o = 13.6$ eV establishes an atomic scale of energies, and this energy E_o is precisely ½ the Coulomb electrostatic energy $U = k_c\, e^2/a_o$ of a proton and electron spaced by the Bohr radius $a_o = 0.0529$ nm. We learned that an atom like hydrogen can generate or absorb photons only when $h\nu = \Delta E = E_o(1/n_1^2 - 1/n_2^2)$, where n_i are quantum numbers of states in the atom. The electron and proton spins have associated magnetic moments, and that the energy difference between parallel and antiparallel arrangements of these moments leads to the 21.3-cm wavelength radiation from hydrogen in outer space and also provides the basis for the cesium atomic clock, through the relation $h\nu = \Delta E$. We learned about the mathematical description of a wave as $y = \sin(kr - \omega t)$. We learned that electrons diffract just like light waves diffract, and the wavelength associated with an electron is $\lambda = h/p$, where again h is Planck's constant. We learned that the propagating property of the matter wave is the wave function $\psi(r,t)$ such that the probability of finding the particle in a given range of space Δr in a given range of time Δt is $P(r,t)\, \Delta r\, \Delta t = |\psi(r,t)|^2\, \Delta r\, \Delta t$.

5.7
What Technologies Are Directly Based on These Fundamental Particles and Spin?

Laser technology has many forms, and the photon is the central player in the laser. Photons come in all colors and beyond the visible range, frequencies that range from 60 kHz (sent from Ft. Collins, CO, USA and from Fukuoka, Japan, to set wristwatches), and also communicate with underwater submarines), to 1.5 GHz timing signals from GPS global positioning satellites, to X-rays used in the dentist's office, in the 10^{19} Hz range. Photons are of more than academic interest! Lasers are used to perform surgery on the retina of the eye. A special "argon excimer laser" generates strong light at 193 nm to expose "photoresist" in Pentium silicon chips, a key aspect in keeping Moore's law on track. Semiconductor injection lasers with wavelength near 1.5 μm are used to light up the worldwide network of optical fibers. Lasers are used to cut steel as well as in grocery store bar-code readers and laser printers. The US military is still spending money trying to make a powerful laser that might provide early, distant destruction of threatening intercontinental ballistic missiles. Lasers are also one focus of an "inertial fusion" effort to make fusion energy on earth, at the National Ignition Facility in the United States.

Magnetic resonance imaging (MRI) and computer hard drives are based on *spin angular momentum* intrinsic to electrons and protons. The MRI apparatus, which images the location of water in biological situations and in people, is based on the spin of the proton. Each water molecule has two protons, and the rules of nanophysics apply to the two states of the proton, up and down with respect to a large magnetic field, is the basis of the imaging in MRI.

A second example of a technology arising from spin is the *"read head" magnetic sensing device* in modern hard disk drives, described in Chapter 2, with more in Chapter 11.

Atomic clocks are based on electron quantum states in atoms. Atomic clocks are based on specific electron states in the cesium atom, which are similar to the states of the hydrogen atom first described by Bohr. The electron states directly involved have specific orientations (parallel and antiparallel) of the electron and nuclear spin momentum in the atoms.

The substructure of the Universe based on particles and new quantum rules allows us to understand clearly the structure of atoms, molecules, and semiconductors. To do this we need to use Schrodinger's equation, which we return to in Chapter 6.

Reference

[1] Dunin-Borkowski, R.E., *et al.* (1998) *Science*, **282**, 1868.

6
Quantum Consequences for the Macroworld

We learned in Chapter 5 that tiny particles of matter are guided by the wavefunction $\psi(x,t)$, which gives the probability $P(x,t)$ of finding the particle at position x and time t as $\psi^2(x,t)$. The question, in a given situation, is how to write Schrodinger's equation, and how to solve the equation.

These new rules do not have much to do with how macroscopic objects, like tennis balls, move, but they do have a huge influence on the composition of bulk matter, for example, the nuclei and electrons that the tennis ball is made of. In some cases the new rules can be exploited to make new devices.

6.1
Quantum Wells and Standing Waves

Schrodinger's equation is simply the statement of constant total energy E of the particle, but requires a new way of describing the momentum: $p = -i\hbar\, d\psi/dx$, where $i^2 = -1$ and $\hbar$ is Planck's constant divided by 2π. The result, since the kinetic energy is $\frac{1}{2} mv^2 = p^2/2m$, is

$$-(\hbar^2/2m)d^2\psi/dx^2 + V(x)\psi(x) = E\psi(x) \qquad (6.1)$$

A free particle ($V = 0$) is described by

$$\psi(x) = \sin kx, \text{ where } k = (2mE/\hbar^2)^{1/2} \qquad (6.2)$$

This is a wave, the wavelength is just $2\pi/k$, seen to depend on the energy E. When the potential V is constant and *larger* than the particle energy E, the solution is a decaying exponential function

$$\psi(x) = \exp(-\kappa x), \text{ where } \kappa = [2m(V-E)/\hbar^2]^{1/2} \qquad (6.3)$$

Understanding the Nanotechnology Revolution, First Edition. Edward L. Wolf, Manasa Medikonda.
© 2012 Wiley-VCH Verlag GmbH & Co. KGaA. Published 2012 by Wiley-VCH Verlag GmbH & Co. KGaA.

6 Quantum Consequences for the Macroworld

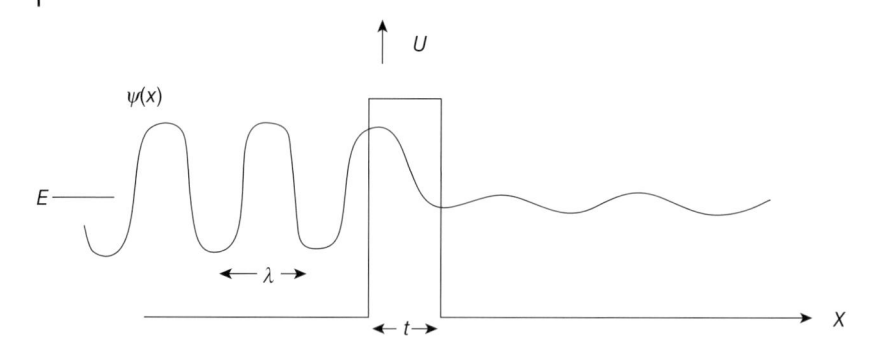

Figure 6.1 A particle moving to the right tunnels through a barrier of thickness t. Note that the energy of the particle, as represented by its wavelength λ, is not reduced. The tunneling probability is $T = \exp(-2\kappa t)$.

This allows a *small* mass to exist in a region of negative total energy. This is not possible in classical physics, nor in common experience! The phenomenon of *barrier penetration* or *tunneling* [1] is based on this, Eq. (6.3). Note that tunneling, as shown in Figure 6.1, would not occur if Planck's constant were zero.

Tunneling is when you hit the tennis ball into the net, and it comes out on the other side! It does not work for tennis balls, but it works for electrons and protons, of small mass. (Tunneling is the basis for the magnetic tunnel junction device [MTJ] used in hard drives, discussed earlier.)

A more interesting and important example of tunneling occurs as shown in Figure 6.2, through a Coulomb repulsion barrier, $U(r) = ke^2/r$, between two protons. This process occurs in the sun,[1] allowing two protons to join together to form a deuteron, the atomic nucleus of deuterium, and releasing energy by fusion [2]. As it occurs on the sun, the tunneling probability is[2] about 10^{-8}. (This is also a prototype for the generation of larger nuclei in stars, by successive tunneling steps at high temperature [3].)

In Figure 6.2, a wavefunction $\psi(r)$ is qualitatively sketched, depicting a proton of energy E coming in from the right. Its wavelength $\lambda = h/[2m(E-U)]^{1/2}$ increases near the turning point r_2 (where $E = U$). The amplitude $\psi(r)$ also increases toward the turning point as the incoming proton slows down. $\psi(r)$ goes into steep exponential decay from turning point leftward to the point of contact. For protons in the core of the sun, with energy $E = 1.293\,\text{keV}$ (at $1.5 \times 10^7\,\text{K}$), $r_2 = 1113\,\text{fm}$, much larger than the spacing needed for contact (fusion) which is $r_1 = 2.4\,\text{fm}$. This process will not occur without tunneling.

The consequences of the fundamental particles and Schrodinger's equation embodying Planck's constant thus include the generation of heat and

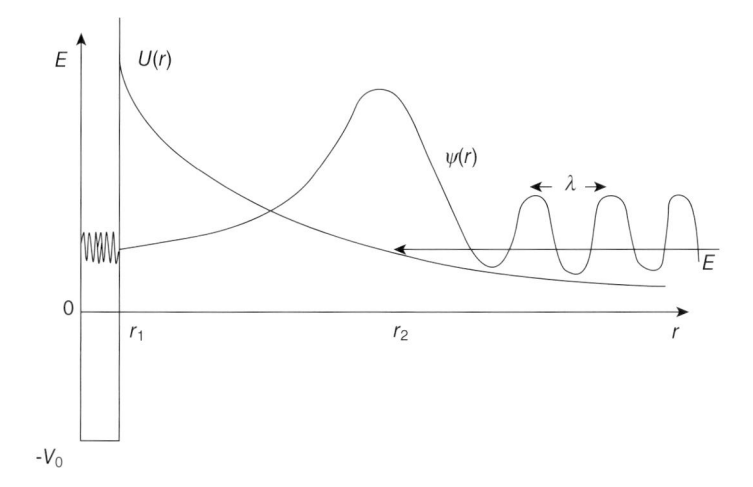

Figure 6.2 Tunneling through a Coulomb barrier, $U = k_e e^2/r$, to form a bound state at $0 \leq r \leq r_1$. A proton approaches from the right indicated by an oscillating wavefunction $\psi(r)$, as it slows down and stops at classical turning point r_2. The tunneling probability for this process [2] for protons on the sun is about 10^{-8}. The rapid oscillation inside the well is because we assume the energy E there is much higher than the local potential V.

generation of the nuclei of chemical elements in stars, these processes all involve the phenomenon of tunneling, based on the wave aspect of particle behavior.

6.2
Probability Distributions and Uncertainty Principle

Solutions to Schrodinger's equation in important cases closely resemble solutions to classical situations of trapped waves. A familiar trapped or standing-wave pattern is the motion of a violin string of length L. The lowest mode of vibration is simply a sine curve, a half-wavelength trapped between two support points, where the displacement y of the string is zero. So $y(x,t) = y_o \sin(\pi x/L) \sin(\omega t)$, $0 \leq x \leq L$, and the maximum displacement occurs for $x = L/2$, where $\sin(\pi/2)$ is 1.0. ω is the frequency in radians per second, for musical note A, $\nu = 440\,\text{Hz}$, and ω is 2765 rad/s.

The same solution form is found for Schrodinger's equation for an electron trapped between two rigid barriers, where the wavefunction $\psi(x) = \sin(kx)$ is the analog of the displacement y of the violin string. An infinite or rigid barrier is a place where the particle cannot exist, so $\psi(0) = 0$ and $\psi(L) = \sin(kL) = 0$. The last condition requires $kL = n\pi$, or $k = n\pi/L$. The

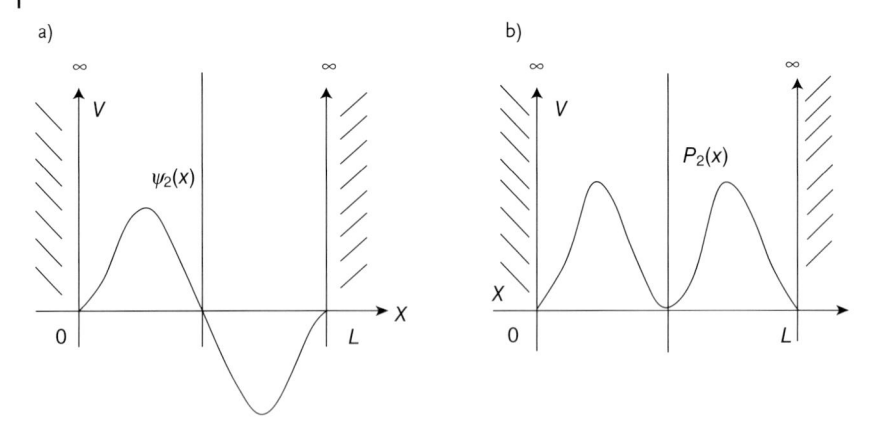

Figure 6.3 (a) Wavefunction $\psi_2(x) = (2/L)^{1/2} \sin(2\pi x/L)$ for $n = 2$ state of trapped particle. (b) Probability distribution, two equal peaks, node at $x = L/2$. The area under the curve is exactly 1 by choice of the normalization.

potential energy inside the "trap" is zero, so the energy of the particle is just its kinetic energy.

The wavefunctions $\psi(x)$ for the allowed states of the trapped electron are similarly just n half-waves trapped between the barriers (since $k = n\pi/L$ implies $n\lambda = 2L$) and

$$\psi_n(x) = (2/L)^{1/2} \sin(n\pi x/L) \qquad (6.4)$$

The probability distribution for Eq. (6.4) has n equal peaks, and the situation for $n = 2$ is shown in Figure 6.3.

Let's find the energy of the trapped particle, which is just kinetic energy $p^2/2m$, where the momentum p is h/λ, according to deBroglie. So, the energy has the square of Planck's constant in it. The energy of the trapped electron is just $p^2/2m$, where $p = h/\lambda$ and λ equals $2L/n$, where L is the spacing of the barriers. Just as in the violin string, the lowest allowed state is ½ wavelength between the barriers, thus $L = \lambda/2$. Using $p = h/\lambda$, we find

$$E = n^2 h^2 / 8mL^2 \qquad (6.5)$$

So the allowed energies are proportional to the square of Planck's constant and to the square of n, the number of half-waves across the trap, and inversely proportional to the square of the trap length L. This is an energy for localization.[3]

These are the exact energies in Schrodinger's equation for the possible states of a trapped or localized particle of mass m in a region of length L. If we put into this formula the characteristic size of an atom, $L = 0.0529$ nm, with $m = m_e$, we get 133.6 eV, a typical electron energy in an atom. If we want to find the trap length L' to match 13.6 eV, it is $L' = 0.166$ nm, about three Bohr radii. Electronvolts and nanometers are convenient units for atomic size systems.[4]

The location of the particle cannot be predicted beyond saying that the probability distribution for finding the particle is $P_n(x) = (2/L)\sin^2(n\pi x/L)$. What can be said is that the particle will never be found at the nodes, just as in the case of the light diffraction pattern. The uncertainty principle, which states that uncertainty in position, Δx, multiplied by uncertainty in momentum Δp comes to less than $h/2$, is satisfied.

6.3
Double Well as Precursor of Molecule

Now let's consider a more complicated 1D case, *the double well*, a precursor to a molecule. Instead of a single well, consider two wells that are joined by a barrier of finite height V_b and thickness t smaller than L. So we have a double well, of total width $2L + t$. This is sketched in Figure 6.4.

Let us start by putting one electron on the left in the $n = 1$, lowest state ψ_A, the single trapped half-wave that we just described.

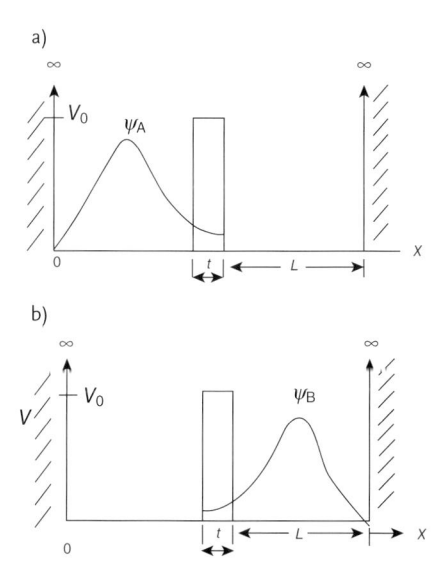

Figure 6.4 Sketch of double well with internal barrier. (a) Electron started on the left in state ψ_A. (b) Electron started on the right in state ψ_B. The barrier in the center is finite, so the particle can tunnel through it, similarly to the process shown in Figure 6.1. These states will transform into one another, rapidly, and this is a way of viewing a covalent one-electron bond.

As before, the value $\psi(0) = 0$, since the left barrier is infinite. However, since the barrier near the center is not infinitely tall, the value $\psi(L)$, at the inner edge of the small barrier is not zero, but has a small value, let's say $\Delta\psi$. The same value must occur just inside the small barrier, because in quantum mechanics the size of ψ cannot jump suddenly but has to be continuous. The result inside the small barrier is $\psi(x) = \Delta\psi \exp(-\kappa x)$ which shows there is a small probability $P(x) = [\Delta\psi \exp(-\kappa x)]^2$ of finding the particle inside the finite barrier of thickness t! We expect that since t is small, then there will is a chance that the electron will tunnel to the right (Figure 6.1). For a barrier of height V_b and thickness t, the tunneling probability is about $T = e^{-2\kappa t}$ as we saw in Eqs. (6.3) and (6.4).

Tunneling may be regarded as the basis for the covalent bond. In Figure 6.4 we see two equivalent starting wavefunctions, ψ_a (on the left) and ψ_b (on the right), each of which will decay by tunneling. *If we have two valid wavefunctions, then an equally valid solution is any linear combination of the two.* This can still describe one electron, but in a linear combination of states.[5] If we consider

$$\Psi_S = 2^{-1/2}(\psi_a + \psi_b) \quad (6.6)$$

this would be two positive half-wave solutions one in each well, giving a totally symmetric wave around the center point of the double well and the $2^{-1/2}$ means that the probability of each case is exactly ½.

We are allowed to choose

$$\Psi_A = 2^{-1/2}(\psi_a - \psi_b) \quad (6.7)$$

This is positive on the left and negative on the right. It is zero at the center (has a node) at the center. This function is antisymmetric, and $\Psi_A(L + t/2) = 0$, which means the electron will never be located at the center.

The symmetric wavefunction will have a lower energy because the electron has a chance of sitting at the center, where it is attracted to both wells at once.

This is a one-dimensional analog of the covalent bond of strength 2.65 eV that occurs in the H_2^+ molecule. As we have said, if the electron is initially on the left ($\Psi = \psi_a$) because of the nonzero wavefunction and tunneling through the barrier, it will end up on the right. More precisely, if started on one side, *it oscillates back and forth between the two portions of the well*. The result is that the binding energy is $h\nu'/2$, where ν' is the rate of tunneling from one side to the other. This is another way of describing the binding of the H_2^+ molecule, the binding occurs by tunneling of the electron back and forth between the two wells. This depends on having h bigger than zero. So molecular formation is also tied up with the value of Planck's h, it must be big enough but not too big!

6.4
The Spherical Atom

The hydrogen atom, which is a prototype for all other atoms, exists in three dimensions and is spherical. Therefore, coordinates x, y, and z don't make sense. The important coordinate is the radius between the particle, (the proton) and the electron. The Schrodinger equation, when expressed in coordinates r, the radius, θ for an angle measured from the vertical direction, the z-axis, and φ, in the xy plane, is rather more complicated, but it still has simple solutions, even if we allow a positive charge Z, instead of $Z = 1$ for hydrogen.

The lowest-energy solution (ground state) is $\psi = \sqrt{(\pi a^3)} \exp(-r/a)$,[6] where a is the Bohr radius, 0.0529 nm. The energy that corresponds to this state, the $n = 1$ state, is $-Z^2$ 13.6 eV, just as in the Bohr model, and, as in Bohr's model, the higher energy states have energy Z^2 13.6 eV/n^2, where $n = 1$ for the ground state. The difference in energy, then, between the $n = 2$ and $n = 1$ states is ¾ of the binding energy, thus 10.2 eV (taking $Z = 1$). *This energy difference is a possible energy received from or given to a photon.*

A new feature of the Schrodinger treatment of hydrogen is that angular momentum values $L = l\hbar$ are allowed for integer values l limited to the range from 0,1,. . . . , n–1. There are historically called "s" states for $l = 0$, "p" for $l = 1$ and "d" for $l = 2$. So the $n = 1$ state is an s-state, with no orbital angular momentum, an error in the Bohr model. For a given value of l, the projection of the angular momentum along the z-direction, (in units of $\hbar$), called m_l, can vary in integer values between $-l$ and l, thus there are $2l+1$ different allowed directions of the angular momentum. For $l - 1$, the allowed values of p_z are -1, 0, and 1 in units of $\hbar$.

These features allow nonspherical wavefunctions and nonspherical probability distributions, leading to directed covalent bonds.

The solutions of the Schrodinger equation for the hydrogen atom in the $n = 2$, $l = \pm 1$ states allow motion of the electron around the vertical z-direction corresponding to angular momentum $L = \pm h/2\pi$. These two states[7] correspond to orbital motions, clockwise and anticlockwise, around the z-axis. These states for $n = 2$ and $l = \pm l$ can be combined to give a *directed wavefunction*, a precursor of the covalent bond. If we add a clockwise motion with a counterclockwise motion we get exactly oscillation along the x-axis.

So this wavefunction, taken as a linear combination of the clockwise and counterclockwise rotational motions around the z-axis, gives us a probability distribution that is like a cigar along the x-axis. If we change the phase we can make a motion that oscillates along the y-axis, a probability distribution like a cigar along the y-axis.

Linear combination wavefunctions are just as valid as the original wavefunctions, in particular they have the same energy. To generalize this, working from the $n = 2$ states of the hydrogen atom, one can construct wavefunctions whose probability distributions point in various directions. One of the most important of these is the set of solutions that point to the corners of a tetrahedron where the nucleus, with positive Z charges, is at the origin. Historically, these solutions are called sp^3 *hybrids*. These sp^3 linear combinations point to the corners of a tetrahedron (Figure 6.5), which describes methane, CH_4, and is also the basis for the structure of the diamond crystal and the silicon crystal. We will talk about these hybrid states again in Chapter 7, in connection with the formation of crystals like silicon.

As we will see, the Schrodinger treatment of the spherical one-electron atom is the basis for atoms with many electrons, and, basically, for all of chemistry.[8] An atom forms quickly around a nucleus as electrons are attracted, up to the point where the atom is electrically neutral, with equal numbers of positive and negative charges.

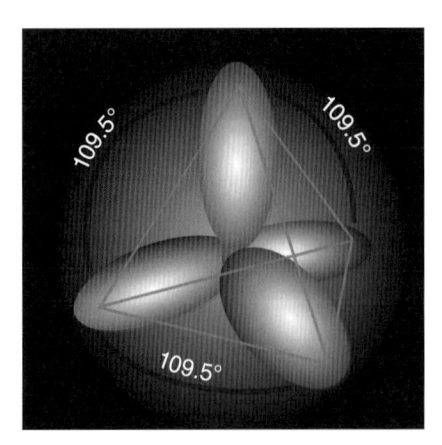

Figure 6.5 Representation of the probability cloud in the tetrahedral hybridized sp^3 states that occur for $n = 2$ in carbon and for $n = 3$ in silicon. The bond angles are 109.5 degrees. These tetrahedrally oriented linear combinations of s and p states appear in the methane molecule CH_4 and the crystal structures of diamond and silicon (http://upload.wikimedia.org/wikipedia/commons/thumb/9/9f/Sp3-Orbital.svg/2000px-Sp3-Orbital.svg.png).

6.5
Where Did the Nuclei Come From (Atoms Quickly Form around Them)?

The building blocks of matter are the elementary particles, which, along with light were formed in the earliest moments of the Universe. We have already discussed the electron and the proton, each having charge $e = 1.6 \times 10^{-19}$ Coulomb and each having spin ½. The next particle is the neutron, which is similar to the proton but has no charge, but still has spin ½. The electrical attraction of the electron and proton confines the electron and forms the hydrogen atom.

6.6
The "Strong Force" Binds Nuclei

The strongest attraction in nature, however, is that which exists between a proton and a neutron (and, in fact, between any pairs among protons and neutrons, including between two protons). The "strong force" or the "nuclear force," which has a short range about 10^{-15} m, leads, in the simplest case, to binding a proton and neutron to form a particle called the deuteron D. (The neutron as a free particle is unstable and decays in about 800 s, but is stable in nuclei.) The deuteron has a binding energy of about 2.24 MeV and so, roughly, its binding energy is 1.6×10^5 times higher than the hydrogen atom. The hydrogen binding energy is 13.6 eV, to take D apart is about 2.24 MeV. The deuteron is the simplest atomic nucleus. The inventory of stable nuclei includes $Z = 92$ for uranium (actually up to $Z = 109$), and the $Z = 92$ uranium nucleus can be stable with several different numbers of neutrons. The nuclear mass $A = Z + N$ protons plus neutrons is 238 for the most common isotope of uranium ^{238}U, but includes 235 for the isotope ^{235}U of uranium that is known to undergo fission, to split apart and to lead to two separate nuclei plus a few extra neutrons, which accommodate all of the nuclear particles. One example of a fission product of uranium ^{235}U is cesium, with 55 protons and 78 neutrons to make $A = 133$. ^{133}Cs is used in the atomic clock which sets the standard second as 9.1923 billion oscillations on this atom. ^{133}Cs is also used in global positioning satellites, in the form of commercial atomic clocks such as the Hewlett Packard 5071A, which are accurate to about 30 ns per day.

The rules of the nuclei determine what atoms can form. The chemical table is really the set of stable nuclei. These are limited by the laws of nanophysics, given the inventory of particles (protons and neutrons), as acted on primarily[9] by the strong nuclear force and the Coulomb force.

6.7
Chemical Elements: Based on Nuclear Stability

The question of what nuclei are stable is separate from the question: how did these nuclei form? Hydrogen atoms from the earliest moments of the Universe eventually formed stars under gravitational attraction, which thus played a central role. As we mentioned in connection with Figure 6.2, protons fuse with tunneling steps at high temperature to make deuterons in stars. A great elaboration[3] of such processes, involving supernova explosions of large stars, is believed to have allowed formation and dispersal of elements, finally to the formation of planets.

Let's consider a few of examples of important nuclei and the corresponding atoms. The deuteron still has nuclear charge $Z = 1$ and so it makes a hydrogen atom with a mass, roughly, twice the mass of a proton. It is called deuterium D and it has a slightly different energy, a very small difference. A much more interesting atom is ^{12}C, which has six protons. For carbon the most common number of neutrons is also six, to produce a weight of $A = Z + N = 12$.

Carbon ^{12}C is the basis for all of organic chemistry and biology. How can we understand the carbon ^{12}C atom? It can be understood approximately within the picture we have mentioned for the hydrogen atom. We can intuitively expect that the number of electrons will be six in order to make a neutral atom based upon the nucleus of charge $Z = 6$. So the six electrons will go into hydrogenic states, the first two will go into the $n = 1$ state, and the spins of these two electrons will be opposite. Four electrons will go into the $n = 2$ state and the spins of these electrons will also add up to zero. The rule that governs this building of an atom is that only one electron is allowed within a completely defined state. This is called the Pauli "exclusion principle," one electron per state, for particles like electrons. The two electrons in the ground state come from the fact that the plus spin and the minus spin are separately counted. The rules of the hydrogen atom are a starting guide to what happens in larger atoms. For example, for $Z = 6$ and $n = 2$, the energy would be $13.6\,Z^2/n^2 = 122.4\,eV$, compared with measured 104 eV. (The close agreement here, neglecting any electron–electron interactions, suggests that the electron–electron interactions are usually unimportant, and we will come back to this in the discussion of a metal.) For $n = 2$, the total number of separately defined states is actually $2n^2 = 8$: of these 2 go into the 2s ($l = 0$) state and 6 into the 2p states ($l = 1$). In carbon, only 4 of these $n = 2$ states are occupied with electrons. So we can go to a larger atom, $Z = 10$ (with 10 electrons) before we need quantum number $n = 3$. This filled shell atom is neon and is regarded as a "rare gas." The atom cesium ^{133}Cs with 55 protons and 78 neutrons is an example of an alkali metal that is one electron past a completely filled Xe core, all states with $n = 4$. So the Cs atom again has a single (now 5s) electron orbiting around a spherical core which has a positive charge near one. This is not a completely correct picture, because the electron can dip into the core and sometimes see a larger charge than one, but it is a good start to understand an alkali metal atom like Cs.

Let's return to our thinking of the carbon atom. A carbon atom is formed because six protons and six neutrons form an especially stable nucleus and this nucleus is naturally surrounded by six electrons to make a neutral atom. This is all described within the presence of the particles in nature with the additional facts, rules of nature, the strong nuclear force binding the nucleus and the Coulomb force binding the electrons to the positively charged nucleus, plus the prescription that only one electron is allowed in a completely described quantum state. This raises the energies as we add electrons to the atom.[10]

6.8
Molecules and Crystals: Metals as Boxes of Free Electrons

There are more subtle interactions that allow the carbon atoms to form crystals, diamond and silicon being the most obvious examples. Diamond is an extended lattice of carbon atoms bound by covalent bonds; see Figure 6.5. Covalent bonds are based on the Coulomb attraction, and the diamond crystal is based on tetrahedral bonding between adjacent carbon atoms. The same is true of silicon, which is perhaps the most important industrial element, certainly from the point of view of electronics and information technology. These crystals, when pure and cool, are insulators: every electron that is present in each atom is bonded to a partner on the next atom, so carbon in the form of diamond, is an insulator.

A metal. A different situation occurs if we have, as in the case of Cs, a completely filled inner core plus one extra 5s electron. Cs forms a wonderful metal, bound by a different kind of interaction than a covalent bond. Let's focus on the description of the electrons in a simple metal like Cs. Remarkably, an empty box of size L (a 3D trap) is a good starting picture for a metal like Cs. To discuss the simple metal we simply add electrons into the states of a 3D trap, that is, the potential is *zero* inside the box of side L, and the potential is infinite outside. We neglect Coulomb repulsion by assuming the accompanying positive ions make the whole system neutral, and the electrons tend to keep out of each other's way.

Our initial discussion of the trapped electron, an analogy to the wave on the violin string, is easily generalized to three dimensions. We think of an empty box along x-, y-, and z-directions of size L and if the particle is confined inside that three-dimensional box with an infinite potential at each wall, the answers are almost exactly the same. The wavefunction will have an integer number of half-wavelengths along each of the directions x, y, and z, and the energy (6.5) that was $h^2n^2/8mL^2$, is now extended to include additional similar energies from the y- and z-directions. So now,

$$E = (h^2/8mL^2)(n_x^2 + n_y^2 + n_z^2) \quad (6.5a)$$

The wavefunction is simply the product of functions, one each in x, y, and z, which is $\psi_{nx,ny,nz}(x,y,z) = (2/L)^{3/2} \sin(n_x \pi x/L) \sin(n_y \pi y/L) \sin(n_z \pi z/L)$. And, similarly to waves on a violin string, these standing waves, are the sum of a right-going wave and a left-going wave, which makes conduction of charge and energy easy to understand.

The electronic state structure is determined really by the building up of the electrons into the allowed levels of the empty box Eq. (6.5a), taken together with the rule that only two electrons can go into a completely described state such as given by specific values n_x, n_y, and n_z. On this simple set of ideas, one finds that the highest filled energy level for an electron system, "the Fermi level," after adding N electrons into the box of side L, to be

$$E_F = (h^2/8m)(3N_e/\pi)^{2/3} = 3h^2/8mL'^2 \quad (6.5b)$$

$N/L^3 = N_e$, and L' is an equivalent length we will mention below. N/L^3 is just the volume electron density N_e, and the numerical value in electronvolts for the Fermi energy gives a value 1.52 eV for Cs metal with a tabulated electron density 8.487×10^{27} m^{-3}. Similar results are useful for important conductors like Ag and Au with Fermi energies of 5.50 and 5.53 eV, respectively. In L' we see the effective size of the box that confines the Fermi level electron, $L' = 1.759/(N_e)^{1/3}$.

The Fermi energy is the most useful parameter to describe a metal, and is accurately calculated from this formula based on the simple idea of the trapped particle.

What causes the binding of a metal? Finally, we note that the binding energy of a metal comes from the reduction in electron kinetic energy related to the delocalization of its wavefunctions! Schrodinger learned of the relation $p = -i\hbar d\psi/dx$ in writing his equation, as we saw in Chapter 5, and the smooth electron states extended away from the atomic cores definitely have lower $d\psi/dx = p$ and therefore lower kinetic energy [$\hbar\, d\psi/dx = p$]$^2/2m$ than localized atomic states. This is the basis for *the metallic bond*. It turns out that metals like tungsten are among the strongest solids.

In this chapter we have learned about Schrodinger's equation, quantum tunneling, and found exact wavefunctions and energies of the trapped electron in one and three dimensions. We learned about the spherical atom as an extension of Bohr's model. We found the principle of linear superposition as the origin of molecular binding, which include hybridized tetrahedral bonds in silicon and diamond. We have learned about the origin of the elements based on strong force, Coulomb force and the Pauli exclusion principle. We have learned that a metal is best characterized by its Fermi energy and that its bonding comes from the reduction in kinetic energy associated with delocalization of its electrons away from atomic states into freely moving states, modeled as free electrons in an empty box.

We will return in Chapter 7 to the industrial semiconductor silicon, to quantum dots and to carbon nanotubes and buckyballs, which are close relatives of graphite.

References

1 Wolf, E.L. (2012) *Principles of Electron Tunneling Spectroscopy*, 2nd edn, Oxford University Press.
2 Wolf, E.L. (2012) *Nanophysics of Solar and Renewable Energy*, Wiley-VCH, Verlag GmbH, Weinheim. See Chapter 2.
3 Rolfs, C.E., and Rodney, W.S. (1988) *Cauldrons in the Cosmos: Nuclear Astrophysics*, University of Chicago Press.

7
Some Natural and Industrial Self-Assembled Nanostructures

In this chapter we will learn about quantum dots, carbon nanotubes, and other nanostructures. To prepare for this, and also to prepare for later topics of injection lasers and the billion-transistor chip, it is important to provide more details as to how silicon conducts electricity. What we say about Si will apply to other important semiconductors such as GaAs and CdSe.

7.1
Periodic Structures: A Simple Model for Electron Bands and Gaps

We have said that pure silicon at low temperature is an insulator. Belonging to the diamond structure, each silicon atom has four outer or valence electrons labeled "p" electrons and "s" electrons. The wavefunctions form linear combinations, called sp^3 hybrids whose probability densities point to corners of tetrahedron. Each silicon atom has four nearest-neighbor silicon atoms arranged in tetrahedral directions. In each case, two electrons, one from each silicon atom, join together to form the *covalent bond*. Thus, every electron in this structure is in a bond with a binding energy ΔE about 1.4 eV. There is a distribution of bond energies, and the smallest is $E_G = 1.1$ eV, which is the bandgap energy. We can see that at low temperature motion of charge cannot occur, if an electric field is applied. Nothing can move; all the electrons are locked in covalent bonds. On this basis, we call the material an insulator.

On the other hand, if we raise the temperature, in thermal equilibrium, there is a chance P given as $P = e^{(-\Delta E/k_B T)}$, where ΔE is the energy to break a bond (and create an electron–hole pair), k_B is Boltzmann's constant, and T is 300 K at room temperature.[1] The dimensionless ratio $\Delta E/k_B T$ determines the chance of an excitation, such as breaking a bond, to occur.

Understanding the Nanotechnology Revolution, First Edition. Edward L. Wolf, Manasa Medikonda.
© 2012 Wiley-VCH Verlag GmbH & Co. KGaA. Published 2012 by Wiley-VCH Verlag GmbH & Co. KGaA.

At room temperature, if we take this excitation possibility into account, there are a small number of broken bonds in the silicon. The broken bond frees an electron into the conduction band at energy $E_C = E_G$ and also introduces a vacancy or hole in the bonding structure, whose energies are in the valence band at and below $E = 0$. Both the electron and the hole can move and conduct electricity. The number of thermally produced carriers in a semiconductor like Si is called n_i, the intrinsic concentration. The number of holes and number of electrons are exactly equal, both come from one broken bond.

We can easily estimate this number for Si from the number of bonds, which is four per atom, and the number of atoms per unit cell which is four, and the dimension of the unit cell, which is $a = 5.43$ Å, so its volume is $a^3 = 1.6 \times 10^{-28}$ m^3 and the chance that each bond is broken, which is $\exp(-\Delta E/2k_B T)$. Here, the $\Delta E/2 \approx 0.7$ eV appears because breaking one bond actually creates two carriers, one electron and one hole. So the result at 300 K, where $k_B T = 0.026$ eV, is

$$n_i = 4 \times 4 \times (1/5.43 \times 10^{-10})^3 \exp(-0.7/0.026) = 6.24 \times 10^{27}/m^3 \times 2.03 \times 10^{-12}$$
$$= 1.3 \times 10^{16} \text{ m}^{-3}$$

This looks large, but in relevant terms it is a small number of free carriers. This number is the product of a very large number of bonds/m^3 (6.24×10^{27} m^{-3}) and a very small probability of breaking ($P = e^{-27} = 2.03 \times 10^{-12}$). Note that only two out of 1 trillion bonds are broken, so this does not weaken the structure! Also, if we evaluate this at 77 K, the temperature of liquid nitrogen, the number of free electrons becomes less than 1/m^3 and the resistivity becomes effectively infinite. So Si is an insulator at low temperatures. The strong temperature dependence makes silicon useful as a thermistor, a temperature sensor.

We need to understand more carefully how the released electron and hole propagate in the silicon, the same discussion will apply when we talk about carbon nanotubes. A central fact about electron propagation in structures such as silicon is that if the electron sees a perfectly periodic potential variation such as a regular of array of atoms in its path, there are magic ranges of energies in which electrons freely move as wave past these atoms without any disturbance. These energy ranges are called *conduction bands*.

The energy range in which this perfect propagation of electrons occurs, the *conduction band* has a starting energy E_C in Si approximately $E_G = 1.1$ eV above the top of the valence band (the energy range of the electrons in the tetrahedral bonds). Similarly, the holes can propagate in the valence band. This band behavior for electrons is not limited to Si, it occurs also for elec-

```
         Nₑ = nᵢ
Eᵤ  ─────────────  e      N_c

        ─ ─ ─ ─ ─ ─ ─   E_F

                        N_v
                  o
         Nₕ = nᵢ
```

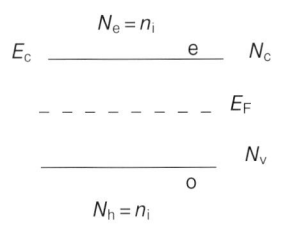

Figure 7.1 Sketch of conduction band and valence band, separated by energy gap of size E_G numerically equal to E_c the conduction band, since electron energy is taken as zero at the valence band. The Fermi energy E_F is midway, for a pure semiconductor. Electron states at energy E above this energy are sparsely occupied according to $\exp[-(E - E_F)/kT]$.

trons in any periodic crystal, for example, in cesium metal and in carbon nanotubes.

7.2
Engineering Electrical Conduction in Tetrahedrally Bonded Semiconductors

Returning to the silicon we see (Figure 7.1) that we have the system in which conduction occurs almost equally by electrons and holes. When we apply an electric field, the motion of the carriers, electrons and holes is described by drift velocity μE that gives the drift velocity in meters per second for the carrier with an electric field E. The parameter μ is called the mobility.

So the average carrier velocity is $v = \mu E$, where E is an applied electric field. The force on the electron is eE, producing an acceleration eE/m and an increasing velocity, $v = (eE/m)t$, which is interrupted by collisions after a mean free time τ. The mean free time τ is related to the mobility as $\mu = (e\,\tau)/m$ (with units $m^2/V\,s$), where e is the electron charge and m is the mass. The collision producing scattering comes with an impurity or by a slightly displaced Si atom in its thermal motion that destroys the periodicity needed for the band motion. The electrical resistivity, ρ, is given by $\rho = 1/ne\mu$, units Ohm m (Ω m), where n is the total carrier concentration in carriers/m^3. If we apply this formula to the intrinsic carrier concentration $1.3 \times 10^{16}\,m^{-3}$ and take the mobility as $0.1\,m^2/V\,s$, reasonable for electrons and for holes in silicon, we find *resistivity* $\rho = 1/[2\,n_i\,e\,\mu] = 2460$ (Ω m). The resistivity is the resistance measured with two electrodes covering opposite faces of a 1-m cube of silicon. This is a high resistivity. The resistance R of a conductor of length L and cross section A is $\rho L/A$.

Nominally the electron mass, 9.1×10^{-31} kg enters the formula $\mu = (e\,\tau)/m$. An effective mass "$m*$," smaller than the free electron mass is found in premium semiconductors such as gallium arsenide. The effective mass can in some cases be as low as 0.01 of the free-electron mass. Since the mass occurs in the denominator in the equation $\mu = e\tau/m$, the electron mobility in these premium semiconductors can as much as 100 times higher. Usually, m^* is given simply as a number, like 0.01 in the case mentioned. This is important in the choice of semiconductors for devices such as quantum dots and field effect transistors (FETs). The chance to enhance the mobility by reducing m^* is important in providing semiconductor devices that act very quickly, i.e., have short switching times.

Now let us return to the properties of silicon as the prototype semiconductor. We have mentioned that thermal excitation will provide us a limited number of free carriers, electrons, and holes in silicon, about 1.3×10^{12} cm^{-3} at room temperature making the material highly resistive. Semiconductor technology is built upon the ability to produce silicon that has *N-type or P-type conduction:* where many electrons or holes are free to move at room temperature. This engineering that is a central part of the nanotechnology in semiconductor devices can be easily understood on an extension of our knowledge of the hydrogen atom.

In the important semiconductors, which include GaAs and Si, have four outer valence electrons, similar to carbon. If we substitute an atom such as phosphorus or arsenic that has five electrons, this atom can be bonded in to the silicon crystal lattice in the same way as the silicon atom. But it leaves one extra electron, that extra electron is weakly bound to the location by the extra positive charge on the dopant ion, P^+ or As^+. This is a less strongly bonded form of the hydrogen atom, in which an electron charge is free to move (now in a conduction band of Si rather than in vacuum) around the positive ion bonded into the Si lattice. The properties of the so-called "hydrogenic donor impurity" can be accurately scaled from the hydrogen atom by correcting for the two main differences from the situation of the proton and the electron in vacuum. These are the higher permittivity κ, which in silicon is increased by a factor $\kappa = 11.8$ from the permittivity of vacuum, and the effective mass m^*. The scaled binding energy 13.6 eV times the ratio m^*/κ^2, where the permittivity κ is 11.8 for silicon and the mass is about 0.3.

The resulting binding energy, 30 meV for silicon, is not strong enough to localize the electron at room temperature. So these doped electrons are released and are free to move in a conduction band and greatly reduce the resistivity. *"Doping" provides a scheme for making silicon of any desired resistivity, either electron or hole conducting.* In fact, silicon can be made essentially metallic by adding donors. In this case there is a range of energies ΔE above the conduction band edge that is filled with mobile electrons, just as in a metal. The formula for ΔE is $E_F = (h^2/8m)(3N_e/\pi)^{2/3}$ just as we found in

Chapter 6 for a metal. Note in this formula that a small m^* can make the energy range ΔE larger. Similarly, silicon can be made heavily hole conducting, as if it were a metal with positive carriers, by adding a P-type impurity like boron, with 3 valence electrons.

Such metallic dopings are used in making contacts, electrical contacts to the silicon in device technology. An FET for example, has the "source" which is a heavily electron conducting electrode, called N^+ and a "drain," which is a heavily hole conducting region called P^+ and these are produced by implanting into the silicon locally high concentrations of a donor such as arsenic or an acceptor such as boron. The basis for the silicon technology is "Electrons and Holes in Semiconductors," the title of the famous book written by William Shockley, one of the inventors of the transistor. The first transistor was not small, but only with miniaturization did this technology become useful. It is now the leading example of nanotechnology.

7.3 Quantum Dots

Let's talk about some of the nanoscale objects that are used in technology.

The first are *quantum dots*, which are tiny crystals of semiconductor (see Figure 3.1). These are used in biological research as color-coded fluorescent markers, and are also built into some semiconductor junction devices. Typical semiconductors for this application are CdSe and CdTe, chosen because their bandgap energies are smaller.

Electrons in a 3D "quantum dot" have a localization energy[2] (Eq. (6.5a)) just as we earlier discussed for electrons in a metal, modeled as electrons in an empty box of side L. The electrons are now in a semiconductor crystal of side L, and exhibit the effective mass for that material.

A "hole" (missing electron) in a full energy band behaves like an electron, except that it has a positive charge, and tends to float to the top of the band. That is, the energy of the hole increases oppositely to the energy of an electron. To create an electron–hole pair in a semiconductor requires an energy at least equal to the energy bandgap, E_g, of the semiconductor.

This application to semiconductor quantum dots requires L in the range of 3–5 nm, the mass m must be interpreted as an effective mass m^* on the order of $0.1\, m_e$. The electron and hole particles are generated by light of energy

$$hc/\lambda = E_{n,\text{electron}} + E_{n,\text{hole}} + E_g \tag{7.1}$$

Here, the first two terms are strongly dependent on particle size L, as L^{-2}, which allows the color of the light to be adjusted by adjusting the particle size. The bandgap energy E_g is the minimum energy to create an electron and a hole in a pure semiconductor. The electron and hole generated by light in a bulk semiconductor may form a bound state along the lines of the Bohr model, described above, called an "exciton." However, as the size of the sample is reduced, the Bohr orbit becomes inappropriate and the states of the particle in the 3D trap provide a correct description of the behavior of quantum dots. So the quantum dot is illuminated with energetic light whose photon energy is bigger than the sum of the three terms $E_{n,electron} + E_{n,hole} + E_g$. Then the electron and hole produced fall down to their lowest energy states, recombine, and give off light, again governed by $hc/\lambda = E_{n,electron} + E_{n,hole} + E_g$. By choice of the semiconductor bandgap E_g and the dimension L of the quantum dot, the color of the emitted light can be adjusted from red to blue.

7.4
Carbon Nanotubes

Recently, smaller self-assembly forms of carbon, mainly based upon graphite, have been discovered and are in the inventory for materials engineering. We will focus on "buckyballs" (carbon C_{60} molecules, ~0.5 nm) and *carbon nanotubes* that are hollow cylindrical shells of carbon, with diameters of a few nanometers. An experimental image of a single-wall carbon nanotube is shown in Figure 7.2.

To understand these useful building blocks of nanotechnology, we have to understand graphite. While we have explained earlier, the diamond structure of silicon and carbon have tetrahedral bonding. Carbon is able to bond in other ways. An "sp^2" bonding allows carbon to form sheets in which each

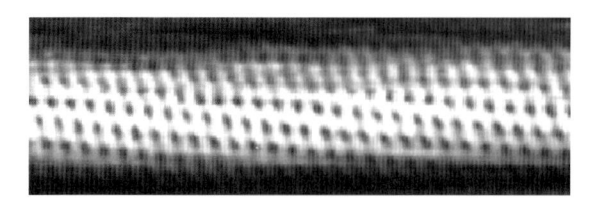

Figure 7.2 An STM (scanning tunneling microscope) image of a single-walled carbon nanotube. Individual sp^2-bonded carbon atoms are shown. This is a "chiral" nanotube because its atoms do not line up perfectly along the axis of the structure. (See the cover of this book.) The carbon–carbon bond length is 0.144 nm (http://www.ncnr.nist.gov/staff/taner/nanotube/types.html).

carbon has three nearest neighbors forming a triangular lattice. This lattice is a sheet in graphite, called *graphene*. Graphite is a layered compound. It is a pure crystal of carbon made up of individual layers that are s–p bonded triangular lattices of carbon. The separate layers are called graphene layers and these have also recently become interesting from the point of view of patterning and nanoelectronics. Graphite forms in nature and probably the best crystals of graphite are mined from the earth. However, the same structure, the triangular lattice of sp^2-bonded carbon appears in smaller structures.

The easiest one of the materials to think about is the carbon nanotube. If we think of the graphene layer, we can imagine rolling this sheet around a variety of crystalline directions. One can imagine a family of cylinders having variable radius and also variable orientation, and all of these are carbon nanotubes. They are found in nature rarely but can be produced in controlled fashions, principally by chemical vapor deposition. The carbon nanotubes with moderately controlled properties, the properties being radius and the index that controls the rolling direction of underlying graphene layer can be made in the following way. Let's imagine a chemical vapor deposition (CVD) system with a heated substrate at ~500 °C and a flow of methane gas CH_4. Controlled decomposition of the methane can deposit atomic carbon on a substrate that is prepared with an array of iron particles of ~1.5 nm radius. These iron particles act as centers or nuclei or catalysts to start the growth of carbon nanotubes.

The way this works is approximately as follows. Iron when it is heated will absorb carbon as an impurity. It readily dissolves carbon and eventually will form a carbide Fe–C compound that has equal numbers of carbon and iron atoms. So carbon can readily be taken up into iron if it is melted to form a ball of radius ~1.5 nm that sits on the substrate surface. The growth that is observed, i.e., that of carbon nanotubes, is approximately 1.5-nm radius growing vertically from such iron catalyst centers. In some cases, the tube would grow up vertically from the iron particle. In other cases the iron particle itself will be raised above the surface by a growing nanotube. An example of this kind of growth is shown in Figure 7.3, of silicon *nanowires*, which are not hollow but filled with silicon.

Controlling the growth of nanotubes is difficult, but dense arrays or "forests" of nanotubes have been grown above a catalyzed surface and the nanotube length can reach a millimeter or more.

It has recently been discovered that nanotubes (however, not made of pure carbon, but of organic compounds) are present in bacteria, which are ancient forms of life, *billions* of years old. (So nanotubes, and other aspects of nanotechnology, are not new!) An example is shown in Figure 7.4, where a

Figure 7.3 Silicon nanowires growing vertically in a chemical vapor deposition (CVD) apparatus at 1100 F. On top of each is a gold nanodroplet saturated with silicon that it intercepts from a silicon-bearing molecule, such as SiF_4, flowing through the reactor. The gold particles, perhaps 10 nm in diameter, were originally deposited on the substrate surface, but are lifted up as the wire grows. This process is similar to one for growing (hollow) carbon nanotubes, but in that case the catalyst particles remain on the surface. ([1]. Markoff) http://www.nytimes.com/2009/09/01/science/01trans.html (Courtesy: IBM).

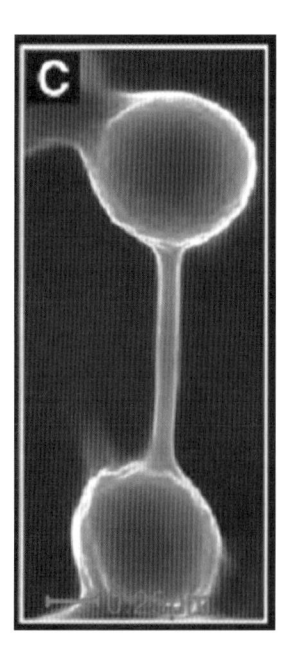

Figure 7.4 Bacterial nanotube. A high-magnification TEM image (×50 000) of a nanotube connecting two *S. aureus* cells. The scale marker is 250 nm. These bacteria appear to be smaller than the magnetic bacterium shown in Chapter 4 (Ref. [2], Figure 6c of Dubey and Ben-Yehuda).

nanotube, made of material similar to the cell wall, actually extends from one bacterium to another. The scale marker is 250 nm. These bacteria are generically similar to those discussed in Chapter 4.

Nanotubes provided in controlled CVD growth, similar to the process shown in Figure 7.3, are single-walled nanotubes, illustrated on the cover and in Figure 7.2. Multiwall carbon nanotubes are also known and are able to telescope, the inner tube fits closely within outer tubes but the interactions between the two are weak and offer little or no opposition to translation or rotation of the inner tube with respect to the outer tube.

What are some uses of carbon nanotubes? The most obvious feature of the carbon nanotube is its very small radius, down to about 1 nm. One application of single-walled carbon nanotubes of small radius is a detector of polar molecules. (Polar molecules have a positively charged region slightly separated from a negatively charged region, to make an electric dipole.) How does this detector work?

If a voltage on the conductor of small radius, electrostatics tells us that the electric field E at the cylindrical conductor is approximately the voltage on the conductor divided by the radius of the tube. Thus, for 1 V applied to a carbon nanotube of radius "a" of 1 nm, the electric field is $1 V/nm = 10^9 V/m$ or a 1000 MVs/m. *This is an extremely high electric field.* It is not strong enough to break up molecules or atoms but it is strong enough to align and attract polar molecules. Many molecules that are commonly used in technology have an electric dipole moment p by virtue of their bonding properties and these dipole moments are tabulated. Molecules that are in the category that we are thinking about include acetone and common solvents, but also includes molecules used in explosives. These molecules with polar electric moments can be pulled out of a gas flowing past the biased nanotube by an electrostatic interaction. This is simply the dipole electric interaction $U = -pE$, which we mentioned in connection with the Van der Waals force. The interaction energy is p (the electric dipole moment) times the E (the electric field). This offers the opportunity to make a small nanotechnology device that can detect dangerous molecules. The numbers show that this can work, what is necessary is that the interaction energy pE be bigger than the thermal energy $k_B T$, at room temperature, which is about 26 meV. This is available at attainable bias voltages and fields with arrays of 1 nm size carbon nanotubes and with dipole electric moments present on molecules of interest. The molecules will form oriented layers around the carbon nanotube. The positively charged carbon nanotube exposed to a gas of polar molecules will attract a layer of oriented polar particles.

How can we detect the attracted layer of polar molecules? The read-out of this detector is by measuring the capacitance $C = Q/V$ of the array of carbon nanotubes with respect to an underlying electrode. The metallic carbon nanotube can be imagined as the inner electrode, radius a, of a cylindrical capacitor. The capacitance of a cylindrical capacitor of length L is $C = 2\pi\varepsilon_o L/\ln(b/a)$, where b is the outer conductor radius and the constant ε_o = permittivity of free space, obtained from $k = 9 \times 10^9 = 1/4\pi\varepsilon_o$. If the nanotube is coated, as we described above, with polar molecules, this slightly changes the capacitance, by changing the ratio b/a. So the readout of the polar molecular detector monitors the capacitance.

This device can be constructed using chemical vapor deposition to make an array of nanotubes. The nanotubes will be deposited on an oxide layer above the silicon surface. The capacitance between the upper layer and the silicon surface is monitored by a capacitance bridge.

Carbon nanotubes are remarkably strong. The Young's modulus is stated as 1 TPa, as stiff as diamond, and they are still flexible, essentially because of their small radii. The sp^2 bonds in the carbon nanotube are among the strongest bonds in nature.

Carbon nanotubes are sometimes added to expensive structural components such as tennis racquets or bicycle frames to improve the strength. Because it is hard to make large quantities of well-controlled carbon nanotubes, these are expensive materials.

As a surprise, it appears that the conditions of an ancient steelmaking process led in fact to carbon nanotube inclusion. "Damascus blades," feared by the opponents of Saladin, the warrior, have been recently shown to include carbon nanotubes. In the process, use of high-carbon steel containing other impurities such as cobalt, favored carbon nanotube growth, which may have made these blades exceptionally flexible and hard.[3]

The electrical properties of carbon nanotubes are quite ideal. The electric current that can be passed through a metallic nanotube can be as high as microamperes, remember that the tube diameter is only a nanometer. The current density is among the highest known in conductors. The nanotube, as examined for example, by an atom-scale microscope (Figure 7.2) is often found to be completely free of defects. It is a perfectly periodic structure and within the understanding of electrical conduction, this will allow electrons to move without scattering.[4] The strong bonds and low weight of carbon make the vibrational frequencies of C atoms very high, which, makes their displacement at given temperature small. So this again favors high conductivity in metallic carbon nanotubes.

It turns out that the nanotube property, the orientation or "chirality," which is hard to control in production, has a big effect on the electrical conductivity. The simple cases are zigzag nanotubes (semiconducting) and armchair (metallic) nanotubes, which are rolled along two principal directions in the graphene sheet. The "band structure" of the nanotube is controlled by this rolling direction.

An alternative approach to finding the right kind of nanotube is to make a lot of them and select those with the right properties. A screening can be done based upon the electrical conductivity of the nanotubes.

But this is really a central problem in exploitation of nanotubes in technology: How to get the right kind of nanotube to the right part of the structure? Steps in this direction have been made making use of DNA technology. As explained in Chapter 4, double-helix DNA forms with "lock-and-key" bondings between bases C to G and A to T between the single strands. If we have a single strand of a given order of bases, it will bond only to a second strand with exactly the complementary order of bases. That is, C on one must face G on the other, and similarly A must face T. The use of this in nanofabrication basically is to take one of the single DNA strands and put it on a nanotube, and put the matching complementary strand on the substrate. If the nanotube with the portion of a single strand is floating around near the substrate, it will bond on to the substrate when the two complementary strands of DNA match up and bind together. This is clearly a high-technology, small-volume kind of operation, but it could be used in organizing carbon nanotubes on a chip in a possible transistor technology. Carbon nanotubes have been demonstrated as conducting channels in field effect transistor configurations. So carbon nanotubes with a source and drain electrode on the ends, and a field electrode (gate) near the tube, act as excellent FETs. Because of its high electrical conduction the ON current of this kind of FET can be high, which is desirable.

The central problem in nanoelectronic application of nanotubes is reliably assembling nanotubes to fill the correct locations on the devices to make a chip.

7.5
C_{60} Buckyball

The related family of sp^2 carbon nanostructures are the buckyballs. The buckyball is a tiny, 60-atom, nearly spherical rolled-up sheet of graphene. Actually there are modifications of the bonding, but mostly sp^2 bonding as in graphene. C_{60} has the same combination of five-sided and six-sided panels as the common soccer ball or football. It is called the buckyball because its structure is the same as the dome built by the architect Buckminster Fuller. These nearly spherical carbon molecules were observed in carbon arc discharges by Richard Smalley and Harold Kroto, for which they shared in the

1996 Nobel Prize in Chemistry. They observed the molecular weights by use of mass spectrometry, where they found a mass equal to 60 carbon atoms. Other magic numbers includes 70. There is a family of "magic numbers" for which "buckminsterfullerene" molecules form with nearly spherical shells. Each of these is a completely reproducible structure. It's about 0.5 nm in size, it is either there or not, it doesn't come in variations. It doesn't have structural defects. It forms exactly as C_{60}. A great deal of effort has been devoted to making use of these remarkable large molecules in electronics. We will mention later a single-electron transistor that was made using a single C_{60} molecule.

In this chapter we have learned about quantum dots, carbon nanotubes, and buckminsterfullerene molecules, making use of introductory information on the electrical conduction of doped and pure silicon. We will make use of this information in the next chapter to talk about injection lasers, which are a basic element in the information revolution, and about the field effect transistor, the basic unit in the Pentium chip.

References

1 Markoff, J. (2009) After the transistor, a leap into the Microcosm. New York Times, September 1, 2009. Image from Dr. Frances Ross, IBM, Yorktown Heights, NY.

2 Dubey, G.P., and Ben-Yehuda, S. (2011) Intercellular nanotubes mediate bacterial communication. *Cell*, **144**, 590.

8
Injection Lasers and Billion-Transistor Chips

8.1
Semiconductor P-N Junction Lasers in the Internet

The Internet information revolution, a key part of understanding nanotechnology, depends almost equally on billion-transistor chips and on fiber-optic communications, which gets your latest video to the cloud computer. A key element in this is the *injection laser* that illuminates the optical fibers. An injection laser is a special type of P-N junction. The P-N junction is also basic in transistors of all types. More about the fabrication of billion-transistor chips, using photolithography, will be included later in Chapter 11, on high-density storage.

As we learned in Chapter 7, a semiconductor is described by a filled valence band, which contains the electron states that form the strong tetrahedral bonding structure. The highest energy in this valence band is taken as the zero of energy. The electron conduction band is located at energy E_G, about 1.1 eV, higher, and this is labeled E_C. In the energy gap, energies zero up to $E_c = 1.1$ eV, no electron states are available. In pure silicon, equal numbers of free electrons and free holes exist by breaking bonds by thermal excitation $k_B T$ according to the probability $P = e^{(-\Delta E/kT)}$. (This was schematically shown in Figure 7.1.)

We found that the number of carriers, electrons equal to holes, called the intrinsic concentration n_i, is ~1.3×10^{16} m³, in Si at 300 K. This is a small number, by an appropriate measure, which is the resistivity ρ, which is very high, ~2460 Ohm m. This resistivity would give a resistance R across a 1-mm cube of silicon as $R = \rho L/A = 2.46$ M Ω. We also learned that by adding phosphorus hydrogenic donor impurity atoms at a concentration N_D, for example, 10^{16}/cc = 10^{22}/m³ we could get almost that same number of free electrons, an increase by a factor of about a million, to get a resistivity ~2.460 mΩ m, so the 1-mm cube would now have a resistance 2.46 Ω.

Understanding the Nanotechnology Revolution, First Edition. Edward L. Wolf, Manasa Medikonda.
© 2012 Wiley-VCH Verlag GmbH & Co. KGaA. Published 2012 by Wiley-VCH Verlag GmbH & Co. KGaA.

```
            N_e ~ N_D
      e e              e  ↓
N_D  ─  ─  ─  ─  ─  ─   E_D
E_F  ─  ─  ─  ─  ─  ─  ─  ↑

           N_h ~ n_i²/n_e
```

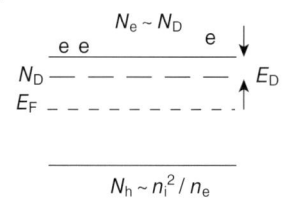

Figure 8.1 N-type case, electron concentration can approach the donor concentration, N_D. Sketch of conduction band and valence band, separated by energy gap E_G numerically equal to E_c the conduction band energy, since electron energy is taken as zero at the valence band. The Fermi energy E_F is now *shifted upward toward conduction band*, to allow higher occupancy of conduction band states, which are occupied according to $\exp[-(E_C - E_F)/k_B T]$.

The band picture appropriate to an N-type case is shown in Figure 8.1.

We learned that the resistivity is the inverse of the product $ne\mu$, where n is the total carrier concentration per m^3, e is the electron charge, and μ is the mobility in units m^2/V s. μ is typically ~0.1 for electrons in Si.

To understand the P-N junction, we need a more accurate description of the occupation of electron and hole states. The Fermi–Dirac distribution function $f(E)$ is defined as the probability that the electron state E is occupied.[1] This still depends only on a ratio $\Delta E/k_B T$, but now $\Delta E = E - E_F$. The Fermi energy E_F, measured from the top of the valence band, can be engineered, adjusted, by doping the material. The Fermi function $f(E,T)$, for low temperature T, is a step function: It is unity for energy less than E_F and zero for energy larger than E_F.[1] A useful approximate value of f far above E_F is $f(E,T) = e^{-[(E-E_F)/k_B T]}$ for $E > E_F$.

For an example in pure Si, assume $E_F = E_G/2$, and let us find the value of $f(E_C)$, at the conduction band edge, $E = E_C = E_G$. Then, $f = \exp(-E_G/2k_B T)$, and with $E_G = 1.1$ eV, $k_B T = 0.026$ eV this is $f = e^{-21} = 6.5 \times 10^{-10}$, nearly zero, for the chance of occupation of a single state. But there are many available states N_C. To find the electron concentration, which we found earlier to be 1.6×10^{16} m^{-3}, f has to be multiplied by the number of available electron states, called N_C. Hence, the band density of states is $N_C = 2.5 \times 10^{25}$ m^3 for Si at 300 K. A similar value is obtained for the valence band N_V.[2]

The chance of having a hole (a missing electron) is $1-f$. At the valence band, $E = 0$, this is a very small number, since f is nearly 1 at $E = 0$ if we assume $E_F = E_G/2$. The chance of a hole being occupied (an electron state being vacant) is again about $f = \exp(-E_G/2kT)$.

This discussion leads us to realize that for a pure semiconductor, with small and equal numbers of holes and electrons, the Fermi energy is at the center of the energy gap $E_F = E_G/2$.[3]

If we have a known electron concentration, because we now know the band density of states N_C we can deduce the Fermi energy (we expect it to be closer to the conduction band edge.) For example, in Si with $N_e = 10^{22}$ el/m^3 we can quickly calculate[4] that $E_c - E_F = 0.144$ eV.

So, the Fermi level moves toward the conduction band if there is N-type doping, and the Fermi level moves toward the valence band if there is P-type doping.

Metallic contacts are obtained in the limit of heavy doping. One can actually make a heavily doped, large-N_D N-type silicon (N^+) contact with many electrons in the conduction band, to make a replica of a metal. In this case, the Fermi energy will be above the conduction band edge by an energy ΔE, using the same formula as calculated for a metal in Chapter 5. So, for the N^+ contact we would have $E_F \approx E_G + (h^2/8m)(3N_e/\pi)^{2/3}$ and an analogous situation occurs with heavy P-type doping.[5]

We now describe a P-N junction, a sharp transition from P-type material on the left and to N-type material on the right. This will have the effect of shifting the bands on the left upward and shifting the bands on the right down, the total energy shift is called eV_B. The voltage V_B called the built-in potential is typically in the range of 0.6 V. The width W[6] over which the shift occurs is called the depletion region. If, for example, $N_A < N_D$, W is mostly in the P-type on the left in the figure, and its width is proportional to $(V_B/N_A)^{1/2}$. Such an "asymmetric junction" is used, for example, in solar cells (Figure 8.2).

How does this junction form, displacing the energy bands? Imagine bringing together a piece of silicon that has a moderate N-doping of phosphorus or arsenic on the right into contact with an identical piece of silicon with a moderate doping of acceptors such as boron on the left. In the N-type region, the Fermi energy lies above the middle of the gap, closer to the conduction band, because we know there are many carriers in the conduction band. Similarly, on the P-side the Fermi energy has to be down toward the valence band because we know there are missing electrons, that is, holes, in that energy range. As we bring the N and P regions together, the Fermi energies on the two sides will move toward to the same value, causing the bands to shift as shown. This is required to get equilibrium: to stop the transfer of electrons from right to left.

Electrons from the phosphorus impurities fall into the boron acceptors states and thereby complete a whole shell of bonding on the boron to make it negatively charged, a space-charge region. The space charge is positive on the N-type side and the negative on the left, the P-type side. This corresponds basically to an electric dipole layer which shifts the bands by an amount V_b. This band shift, V_b, is called the *built-in potential*. The depletion layer width of the P-N junction depends upon the magnitude of the band

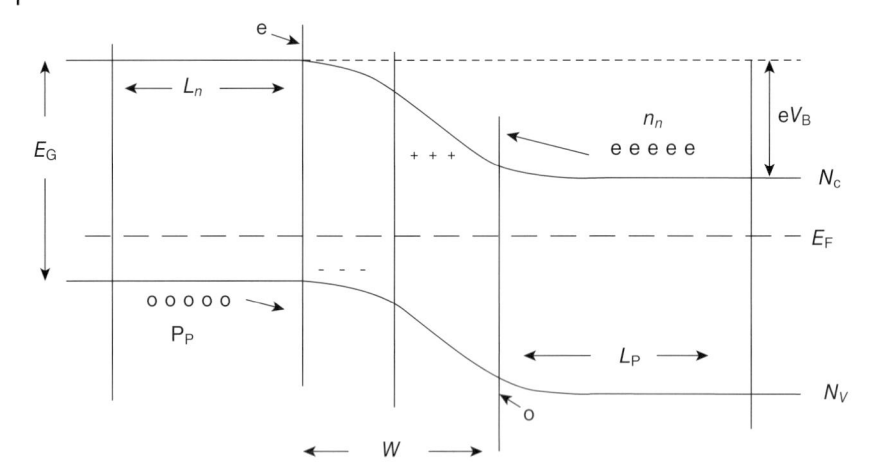

Figure 8.2 P-N junction with band-energy shift eV_B, rising from N-region (right) to P-region (left). In depletion region, W donors (right) lose electrons, to create positive charged region, and acceptors (left) bind the same electrons to create negative space charge. *Minority carriers* (n_p electrons "e" in P-region and p_n holes "o" in N-region) create reverse current density, as they diffuse to the region W where electric field exists. For example, the minority electrons "e" in diffusion length L_n (upper left) fall across the junction in lifetime τ, creating a reverse current density $1/2\, J_{rev} \sim e\,(L_n n_p/\tau)$. Positive applied bias voltage V (not shown) reduces the barrier to $e(V_B - V)$. This will increase the "thermally activated" flow of majority carriers, n_n and p_p, leading to "forward current" $J = J_{rev} \exp(eV/k_B T)$.

bending V_b and also upon the concentrations N_D and N_A of the doping impurities.[6]

A similar barrier, the *Schottky barrier* V_b, occurs at a metal semiconductor contact, for example, a gold or silver contact to N-type silicon. What happens in this case is that the electrons transfer to surface states at the metal semiconductor interface and a completely analogous depletion layer appears with the barrier. Schottky-barrier formation makes producing a low-resistance Ohmic contact to silicon difficult. The Ohmic contact can only be made if the surface is doped heavily with donor or acceptor impurities, so that the Schottky-barrier thickness becomes small and electron tunneling can occur from the metal into the semiconductor.

Let us think about the *I-V* characteristics of the P-N junction device. (This is the most important device in all of silicon technology.) We are going to apply a voltage and measure the current that flows between N-type and P-type regions with a P-N junction as we described. Positive or forward bias for this device is defined as the bias direction that reduces the band shift, and shifts the electron conduction band (on the N-side, the right side) up to allow more ready transfer of electrons into the P-type region, from right to left. We are talking about transfer of electrons from the conduction band in the N-type side to the upward-shifted conduction band on the P-type side, the energy shift being $e(V_b - V)$. So forward bias, positive V, reduces the shift of the bands and increases the current.

At strong forward bias, with applied potential $V = V_B$ the electrons will flow without a barrier from the N-type side to the P-type side and this will be like a short circuit, and will be highly conductive. So the forward I-V characteristics of the P-N junction is governed by the factor $\exp[(V - V_b)/k_B T]$, an exponentially rising characteristic strongly dependent on temperature.

In the reverse bias direction, the current is independent of bias. This current comes from the thermally present (minority) electrons on the p-side falling down the band-bending eV_b into the n-region, and equally from the thermally present (minority) holes on the N-type side falling upward to the valence band on the p-side. Minority carriers have a lifetime τ against being annihilated, as they recombine with majority carriers. The distance, L, that a minority carrier can diffuse in its lifetime is called the "diffusion length." This is the depth (typically much larger than the depletion width W) from which carriers are available to form the reverse current.

The first nanophysical device deliberately engineered in silicon was the Esaki diode, a P-N junction in which the impurity concentrations N_A and N_D were both large so that the depletion barrier was thin and electrons can tunnel between the metallic n-side and the P-type hole states, which are also degenerate or metallic. This produces a distinct I-V characteristic for the Esaki diode, a *hump current* at forward bias in the range of a few tenths of a volt. This produces a negative resistance region that can be used to make an oscillator. Esaki shared the Nobel Prize in Physics in 1973 for this invention.

The P-N junction is the most important semiconductor device. One reason is that it a basic part of all transistors. The junction transistor that was invented by Shockley, Bardeen and Brattain, Nobel Prize for Physics in 1956, can be thought of as two P-N junctions joined together by a narrow base region. So an N-P-N transistor might be two phosphorus-doped silicon regions separated by a boron doped silicon region that is narrow. The device will have three electrodes: emitter, base and collector. The voltage on the base will control the current from the emitter to the collector. These transistors, 3-terminal devices, have led to field effect transistors (FETs) that are used in Pentium chips, as we will describe in a few moments. But we first return to an important form of the P-N junction, again a 2-terminal device.

8.2
P-N Junction and Emission of Light at 1.24 μm

P-N junctions are also used to make lasers. We said above that in high positive (forward) bias on the P-N junction, the barrier potential $V_B - V$ becomes small, and high current density will flow electrons toward the P-type region and holes toward the N-type region. We have assumed so far that all of these carriers will cross the boundary at the boundary opposite side and go around the external circuit. This situation is dominant in silicon P-N junctions, which seldom emit light. The reason is partly that in silicon direct recombination of an electron and a hole giving light, which has negligibly small momentum, is made difficult by a specific feature in the band structure of the silicon.[7]

In "direct bandgap" semiconductors, principally GaAs and its alloys, the electrons and holes in the junction region can easily recombine to emit light. This is illustrated in Figure 8.3, left side.

In gallium arsenide or GaAsP and other direct bandgap semiconductors, the forward-bias current can be primarily turned into light. So instead of having large external current at forward bias, we have a lot of light coming out, primarily of photons at the bandgap energy. The rule for generation of photons, $\Delta E = h\nu = hc/\lambda = 1240\,\text{eV\,nm}/\lambda$ gives a wavelength 1.24 μm for photon energy 1 eV, corresponding to the bandgap of GaAsP. This wavelength propagates well in optical fibers.[8] *The injection laser–optical fiber combination is one of the chief elements of the Internet information explosion that we experience in our use of E-mail, Facebook, YouTube, Google, and similar search engines.*

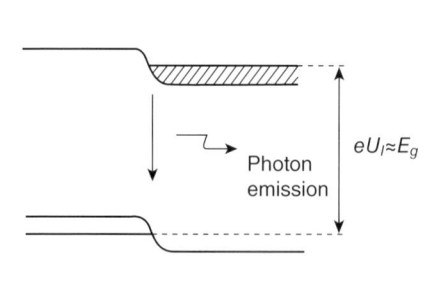

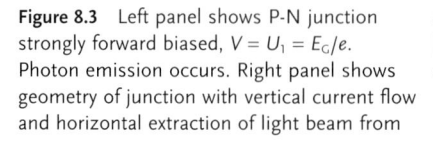

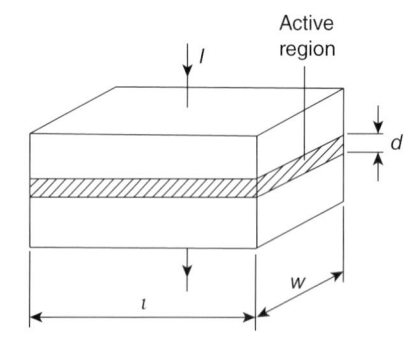

Figure 8.3 Left panel shows P-N junction strongly forward biased, $V = U_1 = E_G/e$. Photon emission occurs. Right panel shows geometry of junction with vertical current flow and horizontal extraction of light beam from active region (Solymar, L., and Walsh, D. [2004]. "Electrical Properties of Materials" 7th edn, [Oxford University Press, Oxford] Figure 12.8, p. 284).

So let's understand more about the injection laser. The forward-biased GaAsP P-N junction allows rapid electron–hole recombination and light emission at the junction. This light is turned into a sharp laser light, with well-defined wavelength, by placing the P-N junction between optical mirrors that select light of particular frequencies and wavelengths, specific *cavity modes*.

Consider a cube 1mm on each side of gallium arsenide that is engineered to have a P-N junction half way down from the top.

Assume the top is N-type gallium arsenide and the bottom is the P-type gallium arsenide. A negative voltage is applied to the top and drives a forward current of electrons down into the P-N junction, and equally a forward current of holes, up into the P-N junction.

At the P-N junction, halfway down the cube, many of these carriers can annihilate, recombine, and make light photons.

Consider a perfect cube of side $L = 1$ mm, let's call the horizontal direction x, where we want the laser light to be emitted, say, into an optical fiber. The edges of the crystal at $\pm L/2$ in the x-direction are perfectly crystalline and serve as mirrors.[9] Thus light of wavelength 1.24 µm is partially reflected and partially leaks out of these boundaries at $\pm L/2$. So this device sends light along the $\pm x$-directions. Because $n = 3.3$ is the ratio c/c', with c' the speed of light in the GaAs, the wavelength of the light in the GaAs is $\lambda' = \lambda/3.3 = 0.376$ µm.

The mirrors make a cavity for light waves, so there are standing waves inside the P-N junction, along the x-direction, analogous to waves on a violin string. The number of harmonics here, however, is very high. The wavelength is 0.376 um and the cavity length $L = 1000$ µm. Thus, there are $N = 2 \times 1000/0.376 = 5319$ half wavelengths for $\lambda' = 0.376$ µm across the x-direction L. The cavity will have a resonance for each number m of half-waves across the length L, and as we have just learned, the mode index values near $m - N - 5319$ will be closest to the desired external laser wavelength 1.24 µm. The photon energy in each mode depends precisely on the index m, the number of nodes across the x-direction of the cavity. Several of these modes near $m = 5319$ (in our example) will have energies that correspond to the light that is emitted by the P-N junction, near 1 eV, the spacing of the modes in external wavelength is $(1/5319)$ 1.24 µm $= 2.3 \times 10^{-4}$ µm. The P-N junction that emits the light, by recombination of electrons and holes, is inside the cavity, and emission is favored, *by the law of stimulated emission,* for energies that exactly match a cavity mode. So the number of photons in the two or three modes closest to the peak of the emission spectrum will accumulate more photons than the other modes. The laser action comes from a rule called *stimulated emission* discovered by Einstein (very shortly after Planck discovered the quantum nature of light in 1900). The rule is that the stimulated emission will go into each mode, which matches the energy, at a rate *proportional to the number of photons in that mode*. So the mode that has the most photons in it is the one that is most likely for the emission of more stimulated photons!

This is the selection process in which there is competition among the electromagnetic modes, the additional energy from the junction funnels into most highly populated mode and this ends in a situation where all of the energy flows into a single mode.

This is called a laser, *light amplification by stimulated emission of radiation*, and the occupation of photons in the main mode becomes very high. It is required that the emitted photon has exactly the same specification (the same energy, the same wavelength, the same polarization) as the mode that does the stimulation. So all of the energy goes into a single electromagnetic mode. And it is this mode that contributes power at one of the boundaries $\pm L/2$ to the optical fiber.

So in summary, we found that by using a direct-bandgap semiconductor such as gallium arsenide to make a light-emitting diode, and to provide it with a cavity with mirrors, it can be easily turned into a laser that provides photons of a single electromagnetic mode, a single energy, a single wavelength, a single frequency and a single direction of propagation.

Lasers are formed in various ways, but the role of the cavity is common among all of them. The means of excitation of the cavity modes differs from situation to situation. Electron injection in a semiconductor laser is one way to produce laser action. Lasers are used widely in technology, from grocery checkout counters, to very high powered lasers used as a cutting tool. Cutting of steel can be done with a high-powered laser.

Before we turn to our next topic, FETs in Pentium chips, we note that laser light is a central tool in semiconductor technology, in the formation of patterns on the silicon surface that can be turned into arrays of transistor devices. Intense sources of extremely short wavelengths are needed, since the wavelength determines the pattern resolution on the chip. We will talk more about the patterning and fabrication in Chapter 11.

In this context, a class of lasers called "excimer lasers" has been developed and is used in the semiconductor industry to produce the 45-nm linewidth class of transistor devices that lead to a billion transistors on a single chip. An *argon fluoride ArF excimer laser* is typically electrically excited, and a photon about 6.4 eV is emitted in the decay of the metastable excited argon fluoride molecule.

Argon Ar and fluorine F ordinarily do not produce a molecule, but on the transient basis in a mixture of Ar and F gases, the ions Ar^+ and F^- can be formed in an electrical discharge. The recombination of the electron from the F⁻ back to the A⁺ gives a strong output of light at about 6.4 eV (thus $\lambda = 193$ nm) and this light is put into a mirror system to form a laser that is used in the semiconductor industry.

8.3
Field Effect Transistor

The FET comes in two varieties, the N-channel FET or N-FET and P-channel FET or P-FET. Both of these devices are built on a single crystal of silicon. Let's talk about the N-FET in which the carriers, when the device is "on," are electrons. The N-FET has (Figure 8.4) highly conducting N-type regions spaced on the surface, these are heavily doped metallic contacts called the source and the drain, and these are the "in" and "out" terminals for the current.

There will be small or zero conduction between source and drain in the absence of a control electrode to bring carriers to the surface. One reason is that the surface of the silicon crystal has a Schottky barrier, that is, the surface has a different electronic structure from the bulk, and the Fermi level of the surface tends to be in the middle of the energy gap. As we learned earlier, pure Si (with the Fermi level in midgap) has only about 10^{16} m^{-3} carriers, which makes it highly resistive. The N-FET device is actually built upon lightly P-type material, so in the absence of any control voltage, the device has few carriers to conduct current from source to drain. The device is off.

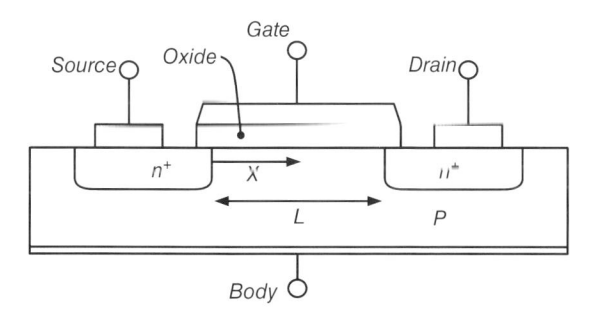

Figure 8.4 Field effect transistor (FET), the workhorse device of silicon technology. This is an N-FET, an electron conducting channel of length L filled with electrons accumulated from the P-type crystal when the gate (top) is biased sufficiently positive. The length L can be reduced in modern devices into the 50-nm range, thus a truly nanotechnological device. This device is reaching the end of its trajectory of miniaturization "the end of Moore's law" (http://upload.wikimedia.org/wikipedia/commons/thumb/7/79/Lateral_mosfet.svg/2000px-Lateral_mosfet.svg.png).

The control element is the gate electrode. The gate electrode in the traditional FET is a metal film deposited on an insulating layer, the gate oxide, which is produced by thermally oxidizing the surface of the Si to produce SiO_2, quartz. The metal gate electrode deposited on the SiO_2 forms a capacitor. The metal is one plate and the surface of the silicon facing the channel is the other electrode of the capacitor. The N-FET device is turned on by biasing the gate positively enough to draw electrons to the surface to form the channel. This is said to be an "inverted layer" because the material itself is P-type and the expected carriers would be holes. But in positive gate bias the gate is sufficiently positive to overcome this and to draw electrons to the surface. Normally, the device would be run with fixed source to drain voltage and variable gate voltage. Gate voltage positive brings up the carrier density in a roughly linear fashion and this is how the control of the current from source to drain is achieved. There is a dependence of the current on the drain voltage, but normally source to drain voltage is kept constant at an optimum value, and control is done by gate voltage.

The device we have described is also called an N-MOS transistor and it has a complementary P-FET form in which the base crystal is N-type, the surface electrodes are heavily doped P-type and the gate electrode draws holes to the surface. It is well known to engineers that making complementary CMOS circuits is fun and productive, so CMOS, complementary metal oxide semiconductor, logic devices are the bread and butter of the industry. These devices are principally used as switches. As the gate voltage is applied, the question is how quickly the device will turn on. This depends upon several factors, including the RC time constant for the gate electrode capacitor, dependent on its size, on the length of the channel, in which case smaller is definitely better, and on the mobility of the carriers. Mobility is a materials property, dependent on the effective mass of the carriers. These properties can be manipulated also by introducing strain, in an engineering fashion, in the channel region. The fact is that the present devices operate at at least 1.8 GHz, which implies a time of 0.55 ns.

Improvements in device count and in device operating speed have been achieved by *scaling*, making the devices smaller in a reasoned fashion. The silicon parameters do not change, the depletion width and the properties of the contact do not change, but the parameters that have been changed are the length of the channel or the gate length, the lateral dimensions of the overall device, and the thickness of the gate oxide. The scaling has had the effect of reducing the oxide thickness as well as reducing the channel length.

The present situation of the scaling of the FET is that the gate length is about 50 nm, and the carrier transport along the channel is nearly ballistic.

This means that the carriers go from source to drain with only one or two scattering events.

The gate design has been changed to avoid electron tunneling leakage across the gate. This was a problem that was solved in the past few years. The scaling had brought the SiO_2 thickness down to only a few atom thicknesses, as was required to maintain sufficient gate capacitance to turn the device on with the available gate voltage. The cure was to maintain gate capacitance without reducing the oxide thickness, by changing the oxide to one of higher permittivity κ. The capacitance of a plate capacitor is $C = \kappa \varepsilon_o A/t$, where t is the thickness. If κ is increased in the same proportion as t then the capacitance remains unchanged. So what was done was to increase κ from about 6 for the SiO_2 to about 20 by choice of HfO_2, making the barrier thick enough to avoid tunneling. The challenge, then overcome, was to learn how to make excellent epitaxial layers of HfO_2 on the surface of Si. This was done by a method called *atomic-layer deposition*. This is a variation of molecular beam epitaxy that produces epitaxial interfaces, that is, the atoms line up on both sides of the interface.

Atomic-layer deposition, now a part of chip production, and definitely an element of nanotechnology, is done by a two-step process, two steps for each molecular layer of the desired material, let us say HfO_2. The overall growth requirement is that an epitaxial relation exists between the substrate, which might be Si [100], and the oxide to be grown, which might be HfO_2. Assuming this requirement is met, atomic layer deposition is an iterated AB process. Step A is to bring in the metal (in this case hafnium), and allow it to form an atomically perfect monolayer in registry with the Si exposed surface. This is accomplished with a Hf-containing molecule, sufficiently volatile (i.e., to have a large vapor pressure at a practical furnace temperature), to flow in as a gas that is then decomposed on the Si surface. Such a volatile molecule is $HfCl_4$, hafnium tetrachloride, which is carried to the substrate in a carrier gas such as argon. The $HfCl_4$ is decomposed by high temperature at the substrate, and the Hf atoms fall into their sites until that set of sites is filled, and the released chlorine flows away. In step B the flow of gas is changed to a flow of oxygen or water vapor that will fully oxidize the Hf layer, leading to one monolayer of HfO_2. This completes the 1st monolayer, and the assumption is that this layer of HfO_2 will nucleate the next layer of Hf atoms, from decomposition of $HfCl_4$. This (A,B), (A,B) process is repeated to grow the film, the total thickness can be completely controlled as integer numbers of the monolayer thickness. This must be a successful process because it is used in the "45-nm node" line of state-of-the-art silicon chips.

In this chapter we have learned about the Fermi function, electron and hole concentrations in pure and doped semiconductors, and P-N junctions and their use in Esaki tunneling diodes and in injection lasers used to generate light for the fiber optics that span the earth. We have learned also about FETs of N-FET and P-FET varieties, that are used in computer chips. We have learned about the argon fluoride excimer laser that creates 193-nm light used in patterning chips that contain a billion FET units. We have learned about the high-kappa dielectrics that are used as gate insulators in current "45-nm node" transistors and a bit about the atomic layer deposition process that was developed to epitaxially deposit the HfO_2 insulating layers. In Chapter 11, devoted to high-density storage, we will describe more details of the photolithographic patterning that is essential to making billion-transistor chips.

9
The Scanning Tunneling Microscope and Scanning Tunneling Microscope Revolution

The scanning tunneling microscope or STM is a scanning device that is able to provide images of single atoms and even single covalent bonds, which are directed electron probability clouds, as we have described in Chapter 5. The spatial resolution of the STM device is on the scale of the Bohr radius, which we found in Chapter 5 is 0.053 nm. It is useful for studying flat conducting surfaces and observing and even manipulating single atoms and molecules that are deposited on such surfaces. It has led to a larger family of scanning probe microscopes, notably the atomic force microscope (AFM) and also the possibility of dense data storage using tips to write and to read data. We will return to the latter topic in Chapter 11.

9.1
Scanning Tunneling Microscope (STM) as Prototype

The scanning tunneling microscope (STM) was announced by G. Binnig and H. Rohrer in 1982. The basic elements of their invention, as shown in Figure 9.1, are a metal tip that is moved in three dimensions x, y, z by a control system that measures the tunneling current between the tip and surface and uses a feedback or servosystem to maintain a constant current as the tip is scanned across the surface.

Motion of the tip is provided by piezoelectric elements p_x, p_y, p_z. A piezoelectric element, as we mentioned in Chapter 3, with regard to the quartz clock in the personal computer, is an element whose dimension changes with electric field, and the electric field is applied by voltages on metal electrodes on the piezo faces. Nanometer-scale movements are easily obtained with voltages in the 1–100 V range. The motion of the tip is carried out by a control unit (CU) by means of piezovoltages such as V_p that controls the

9 The Scanning Tunneling Microscope and Scanning Tunneling Microscope Revolution

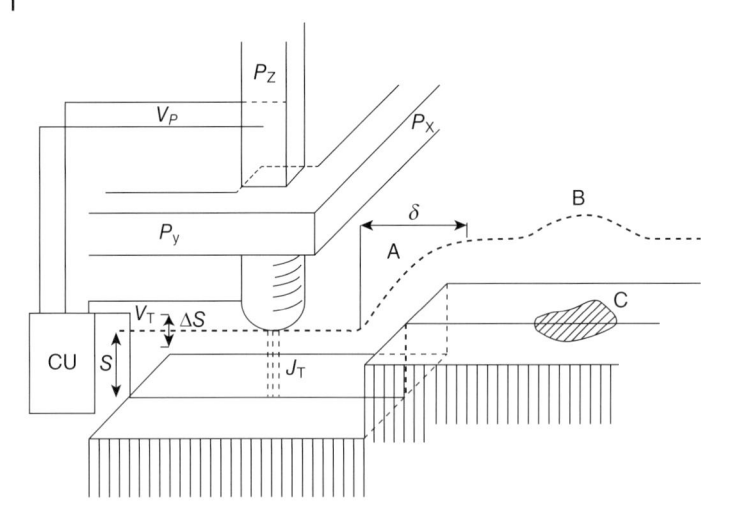

Figure 9.1 Original depiction of the STM by its inventors, Gerd Binnig and H. Rohrer of the IBM Zurich Laboratory. A tip is positioned and biased at voltage V_T to provide a tunnel current J_T to the surface of interest. A control circuit moves the tip up and down to maintain constant current, as the x-position is scanned, to reveal, in the z-piezo voltage V_p, an indication of the step height that is overcome. The region marked C illustrates that a false height can result from a change in surface condition. The device has proved of immense importance. This is a nanotechnology device by virtue of the quantum tunneling process and the 0.1-nm-size scales it reveals (Binnig, G., Rohrer, H., Gerber, Ch. and Weibel, E. [1982]. Phys. Rev. Lett. 49, 57).

tip–surface spacing, t. The vacuum between the tip and the sample acts as a tunneling barrier of thickness t, as we described in Chapter 6.

The device is operated in ultrahigh vacuum and requires atomically clean scanning tips and sample surfaces. Control over the relative positions of sample and tip was the great achievement of Binnig and Rohrer, who made adroit use of piezoelectric control elements in three dimensions coupled with feedback circuitry. The tip is usually a wire of tungsten or iridium etched to a sharp point, but silicon tips have also been used.

The tunneling current between tip and sample at any fixed bias voltage is a rapidly decreasing function of the separation t, basically because the electron probability density outside a metal falls off exponentially, approximately as $\exp(-4\pi(2m\varphi)^{1/2}t/h)$ where h is Planck's constant and φ the work function, closely related to material in Chapter 6, see Figure 6.1. The variation can also be thought of as $\exp[-t/a^*]$ where a^* is an effective electron orbital radius, perhaps 0.1 nm. The effective orbital radius is dependent on the metal work function, φ, which is the barrier height for electrons leaving a metal. Values of the work function are in the range of 2–5 eV. Maintaining constant current as the tip is slowly scanned across the surface amounts to maintaining a constant separation *t*, and thus a *topograph* is generated. Binnig and Rohrer demonstrated their invention by displaying a detailed image of the complex 7×7 reconstruction of the Si [111] surface,[1] a great achievement and the opening of a new era in surface science, and earning, as well, as a share of Nobel Prize in Physics in 1986. From a technical side, the success in atomic-scale imaging demonstrated that piezoelectric elements, under smooth changes in bias voltage, can provide smooth motions on an Angstrom scale, free of jumps.

The scanning tunneling microscope (STM) invented by Binnig and Rohrer often achieves atomic resolution, an unexpected result in view of the relatively blunt apex of an etched tungsten tip. The basic reason is the extremely rapid falloff or decay of the electron density outside the tip, a decay distance on the order of 0.1 nm. Even if the tip is an etched tungsten or iridium wire, achieving a radius on the order of 100 nm, it has some roughness, and in the end a single atom may protrude enough to dominate the tunneling current. In practice, the tip is often decorated with a single atom, which can even be picked up from the sample surface by appropriate manipulation of tip voltage. The tip can be used for nanofabrication that in its simplest form is moving single atoms on a metal surface. One of the most famous images is the STM picture of 48 Fe atoms arranged in a circle, a "quantum corral," achieved by D.M. Eigler, a physicist at the IBM laboratory in Almaden, California. The process of sliding atoms along a surface was called "nudging," by Eigler and his collaborators (Figure 9.2).

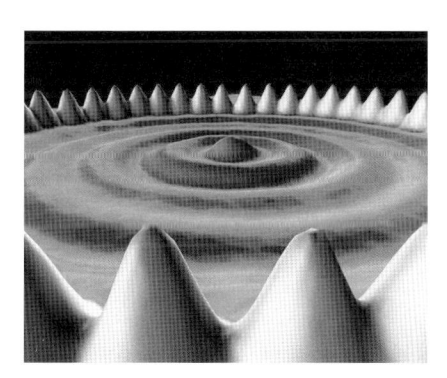

Figure 9.2 View of 48 iron atoms arranged in a circle on a Cu surface, obtained by STM in the laboratory of Donald M. Eigler at IBM in Almaden CA. This image is based on data, but the perspective view shows how computer programs can be used to enhance features. The ripples in this picture are direct evidence of the wave nature of the electrons in the surface layer enclosed by the Fe atoms. The ring of atoms was formed, after an evaporation of Fe onto a clean Cu surface, by moving the Fe atoms one-by-one using the STM tip. Nanofabrication is thus achieved but it is a very slow process.

In some cases an atom or molecule can be picked up, for example, a study of magnetic atoms on a copper surface shows nanophysical effects from a phenomenon of "Kondo scattering" from a single magnetic atom. These workers started by making a carefully etched gold tip. They then transformed the gold tip by carefully inserting it into the clean copper substrate surface to form a uniform layer of copper atoms coating a gold tip. Then the measurements were made between this engineered copper tip and a copper surface that had a few cobalt atoms on it. So the engineering of atomic scale tips has frequently been achieved. Carbon monoxide molecules and even C_{60} molecules have been picked up on conventional tungsten or iridium tips, and once this has occurred the resolution of images is greatly improved.

As an additional example of nanofabrication by the scanning tunneling microscope, chemical reactions have been carried out step by step on single-crystal surfaces of Cu.

In this case, organic chemistry was carried using an STM tip on a copper surface. The starting molecules are iodine-substituted benzene rings, known as iodobenzene. Benzene, C_6H_6, has a pure hexagonal structure and is bound together by carbon–carbon bonds built from sp^2, linear combinations of 2s and 2p wavefunctions that are planar. The reaction, known as the Ullman reaction, is the copper-catalyzed transformation of iodobenzene C_6H_5I to make $C_{12}H_{10}$, an important chemical known as biphenyl. This reaction is known as the Ullman reaction. The STM reaction was carried out at 20 K, with the usual thermal activation energy replaced by pulses of energy from the tip. The first pulse from the STM tip was applied to dissociate the I, from the iodobenzene, leaving a phenyl group C_6H_5. A second voltage pulse is applied to cause two adjacent phenyl molecules to bind into $C_{12}H_{10}$.

In the STM experiment the iodobenzene molecules were lined up along a monoatomic step on the surface, which offers stronger bonding than does a flat surface. In the first action by the tip, a pulse breaks off the iodine, leaving a phenyl ring. The STM images the results at each step. The tip is then used to pull the iodine out of the way, and to pull two phenyl rings closer together. The second tip action provides energy to allow the two adjacent phenyl rings to fuse and form $C_{12}H_{10}$, biphenyl, the desired product. Finally, pulling on one end of the biphenyl was demonstrated to move the whole molecule.

This experiment shows a new use of the STM tip, in adding energy to either dissociate a molecule into parts or to help two molecules to react, by providing energy to overcome a barrier to the reaction. In this experiment a prepared Cu surface was employed, one cut at a slight angle to produce monoatomic steps. These locations offer stronger binding to an atom or molecule, because two sides of the atom are attracted, rather than one, to a flat surface.

Other aspects of scanning tunneling microscope imaging and manipulation of molecules are demonstrated in the deposition and imaging of acetylene, C_2H_2, on the Cu (100) surface at 8 K. Acetylene appears in an STM image as a dumbbell-shaped depression on the surface with a maximum depth of 0.23 Å. This means that the acetylene molecule, in effect, makes the tunneling barrier thicker, so the tip has to go closer to the surface to keep the current constant.

Two experiments were carried out. In the first, starting from a stable clean metal tip, a single acetylene molecule was transferred to the tip, by positioning directly above, and applying a voltage pulse of 0.6 V, 100 nA, for 1 s. After this, atomic resolution was obtained, which had not been available for the metal tip before picking up the acetylene molecule. Then the imaged area, 2.5 nm square, was scanned at a sample bias of 100 mV and tunneling current 10 nA showed the copper atom spacing as 2.55 Å and a corrugation (height modulation) of 0.009 Å.

In a second experiment an example of tunneling spectroscopy was carried out. We earlier mentioned that diatomic molecules have vibrational frequencies v in the vicinity of 10^{14} Hz. The STM can be used to measure the vibrational frequency, through a small abrupt change in the slope dI/dV at the threshold energy, $eV = hv$. In the present experiment, the metal tip was placed directly above the acetylene molecule, and the tunneling I-V curve was very carefully measured. The curve showed a small change in slope dI/dV at the energy $eV = hv = 0.32$ eV, corresponding to a vibrational frequency 7.7×10^{13} Hz.

Returning to fabrication methods for tips, the smallest tips that have been nanofabricated in large numbers are Si tips that were used in the IBM "Millipede" high density storage device. This device, which we discuss in Chapter 11, has an array of 1024 tips with final radius 20 nm. These were produced on a silicon surface by the photolithographic technology used in the semiconductor industry. Pillars of silicon were produced by etching the silicon surface that had been prepared with a pattern of "photoresist," a thin chemical layer that will protect a surface against etching. These pillars are then manipulated in a process called oxidation sharpening to produce sharp silicon tips of radius 20 nm. The tip was used to make a nanometer-scale dimple or indentation in a polymer surface, a means of recording a single bit of information. This allows a chance of data densely encoded on a surface. The details of forming the tips at the end of cantilever springs made of silicon are described in Chapter 11 on dense data storage.

9.2
Atomic Force Microscope (AFM) and Magnetic Force Microscope (MFM)

The atomic force microscope (AFM) as sketched in Figure 9.3, is an important scanning probe tool for nonconducting surfaces. This device is based on a tip mounted upon a cantilever or a spring. We discussed the resonant frequency of a cantilever in Chapter 3, where we learned that the frequency depends on the length L as $1/L^2$ and is also depends on the stiffness Y and mass density ρ of the material.

One of the standard modes of atomic force microscope is to scan the cantilever across the surface and create a topograph by maintaining *constant deflection* of the cantilever. The tip at large spacing will feel an attractive force from the surface because of the van der Waals attraction, an effect we will describe in a moment. If the tip comes closer the interaction will become repulsive and the tip will eventually run into the surface and deflect strongly.

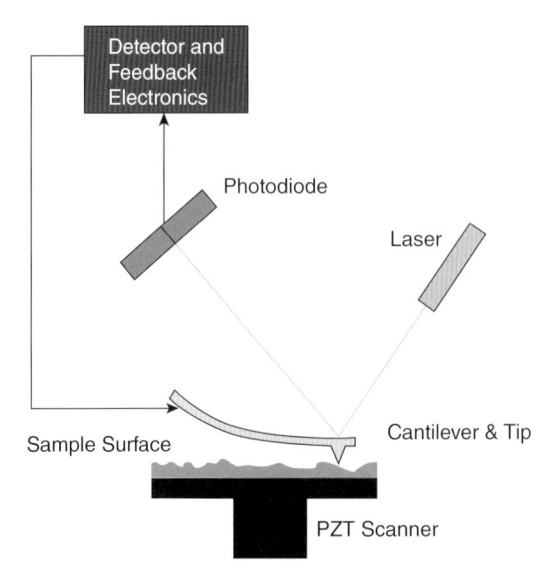

Figure 9.3 Split-field photodiode, with laser and mirror, as a detector of AFM cantilever deflection. The cantilever is carried by a piezodrive (not shown) that scans laterally and adjusts height to maintain constant deflection and thus constant force. This diagram illustrates the notion of hybrid technologies as mentioned in Chapter 1. The laser and the photodiode (basically a P-N junction) as well as the miniaturized cantilever and possibly atomically sharp tip, are all nanotechnology, or have been aided by some aspect of nanotechnology (http://en.wikipedia.org/wiki/File:Atomic_force_microscope_block_diagram.svg).

There are several modes of operation of the atomic force microscope. In the initial devices, the cantilever deflection is monitored and a constant deflection is maintained by pulling the tip up or down, using piezoelectric elements. The control or "servo" loop is similar to that for the scanning tunneling microscope (Figure 9.1).

The initial method for sensing the deflection of the cantilever was based on reflection of light, typically a laser light beam from the upper surface of the cantilever, which acts as a mirror. The light was reflected onto the face of a (PIN) photodiode, whose face was divided into two regions. The circuitry determines which half of the photodiode is illuminated, and the servo loop (feedback circuit), via the piezovoltages V_p, adjusts the tip height to maintain equal amounts of light onto two portions of the PIN diode. This ensures constant deflection of the cantilever, which means constant force on the tip. The AFM tip is at the end of the cantilever, and the cantilever–tip assembly is mounted on a piezoelectric suspension as in the STM (Figure 9.1). The tip is small and sharp to promote high resolution of surface features, and the cantilever[2] is small and lightweight to allow a high resonant frequency, as was discussed in Chapter 3.

Let us think about the force between the sample and the tip that occurs when the spacing decreases. Which way will the tip be deflected? Contrary to one's initial thought, the force is actually attractive, when the tip is not too close.

Let's talk about the *attractive force* that exists when the tip is distant from the sample, which is called the van der Waals force. This force is used in one of the standard modes of operation of atomic force microscope. The van der Waals attraction occurs between neutral pieces of matter, such forces are important between biological molecules like proteins.

Let's make a simple discussion. Let's think of *two hydrogen atoms*. One will be the tip and one will be the sample, quite a simplification! In the hydrogen atom, the electron jumps or hops at high rate among random positions, mostly about one Bohr radius from the proton. The distribution of radius values is given by a function $P(r) = 4\pi r^2 e^{-r/a}$, where a_o is the Bohr radius. The random jumping motion is very different from a regular orbit. At every instant in this jumping motion, the electron and the proton form an electric dipole $p = ea$, i.e a positive and negative charge displaced in a particular direction by a distance about one Bohr radius.[3] A dipole $p = ea$ creates an external electric field $E(r)$, axially symmetric around the dipole direction, and decays as $\sim r^{-3}$.

Because of the jumping electron motion, this electric dipole rapidly fluctuates in its direction, producing a fluctuating electric field, nearby and in particular at the location of the second hydrogen atom.

What effect does this random jumping electric field have on the second hydrogen atom? (Remember we are thinking of one hydrogen atom as the AFM tip and the second as the surface.)

The answer is that there is an attraction, the van der Waals attraction, of the two atoms. The electric dipole field of the first atom will polarize the second hydrogen atom at radius r to create an *induced or replica* dipole on the second atom, which we can call $p_i = \alpha\, E(r)$. α is the polarizability of the hydrogen atom, which is a well-known quantity. And so, we have this consequence, an *induced* dipole p_i on the second atom, which points along the direction of the electric field from the first atom. In energy terms, the interaction U of the induced electric dipole with the driving electric field is $U = -p_i\, E$, where p_i is the induced dipole and $E(r)$ is the electric field generated by the first atom. This pulls the two atoms together.

Both of these quantities scale as $1/r^3$, so their product, the attractive energy, scales as $1/r^6$. This is the characteristic of a van der Waals attraction.

The induced dipole, which is $p_i = \alpha\, p_0/r^3$ parallel to the original field that also is proportional to $1/r^3$, finds an attractive energy proportional to $\alpha p^2/r^6$. Now this $1/r^6$ is the characteristic spacing dependence of the attractive van der Waals interaction. It is an attractive interaction that occurs between all bits of neutral matter. It is much weaker than the Coulomb interaction and also falls off more rapidly with spacing. But the assumption is that in the AFM situation we are discussing, both atoms are neutral, so there is no Coulomb force. So, the van der Waals force is the most important force. (We assume that the two electrons in the two hydrogen atoms are far enough apart that they don't form a covalent bond. A covalent bond would occur only once the hydrogen atoms are spaced on the order of the Bohr radius.) Since this attractive energy is proportional to $1/r^6$ where r is the spacing of the two atoms, the resulting force varies as $1/r^7$.[4] This force will be felt by every atom in the tip interacting with every atom in the surface.

As a result there will be an attractive force, strongly dependent on spacing between the AFM tip and the surface to be imaged, and this is the basis of the AFM operated in the attractive mode.

In this mode of operation of the AFM, the tip and sample are fairly far apart and so the spatial resolution is less than if the tip is very close, as in a repulsive mode. So the tip deflects toward the sample, and the spacing is adjusted to keep that deflection constant.

In a repulsive mode, where the tip may actually come into the repulsive range of interaction between two atoms, one can expect a resolution that is more of the sizes of atoms, although it will also be related to the sharpness and size of the tip.

Now, we can describe the mode of operation of a second form of AFM. The device is the same except that the cantilever is put into oscillation. The cantilever in this case will be made of a piezoelectric material, such as quartz, so that the signal from an oscillator. For example, one might imagine the AFM tip in Figure 9.3 as attached to the Bulova cantilever shown in Figure 3.3. A feedback signal can put the cantilever into oscillation and its oscillation frequency will be determined by the effective spring constant and mass of cantilever. So, this is a mass on a spring situation, it will have a very well-defined resonance frequency. A cantilever will be designed so it has a high Q and a very sharp resonance, and the feedback loop will provide a positive feedback to keep the oscillation going. And a frequency measuring device, a counter, will monitor the frequency of the cantilever's continuing oscillation. When the cantilever gets close to the surface and the attractive van der Waals interaction comes into play, it effectively alters the spring constant (or provides an additional spring between the tip and the surface). As you can see, altering a spring constant or making an addition to it will change the frequency of oscillation. Since the oscillation has a very well-defined frequency, determined as $f = (1/2\pi) (K'/m)^{1/2}$, with K' the total spring constant, the resulting shift in frequency can be detected. So, the mode of operation of the modern AFM is to scan the tip over the surface and keeping the height adjusted to maintain a *uniform shift in the frequency of a cantilever*. This mode keeps the tip away from the surface, which minimizes the chance that there is any damage to the surface or the tip.

There are several modifications in the operation of the modern AFM, which is an important tool in materials and biological science. One modification is to use the mode that we have just discussed to keep the sample and tip at fairly large spacing and on a pixel basis stop the scanning motion and allow the tip to come closer to the sample, where it can measure, point by point, the repulsive interaction, which we have said, offers a higher degree of spatial resolution. This would be a hybrid mode of operation, which can be called a "tapping mode" that is used in certain applications in atomic force microscope.

Modifications and extensions of AFM are extensive. One modification is the magnetic force microscope (MFM) in which the tip has a small bar magnet attached to it. This can be used to determine the distribution of magnetic field emanating or sticking up from the surface. A magnetic field B will emanate from a ferromagnetic domain (like that on a hard disk) in which case the magnetic force microscope will deflect up for one domain direction and down for opposite domain direction and would easily show the location of domain boundaries. If such a microscope was scanned across the data element on the hard disk drive, which a modern disk drive is an array of tiny magnets polarized up or down with respect to the surface, the magnetic force microscope would record points – positive or negative peaks. The MFM devices are well developed and commercially available.

A research device is a further collaboration on the magnetic force microscope and this is called the magnetic resonance force microscope (MRFM) that has been demonstrated as able to detect single-electron magnetic moments in a sample surface.

Scanning probe methods for investigating magnetic field distributions $B(x,y)$ include two other important scanning probe devices. The simplest is the Scanning Hall probe microscope. The sensing element is a highly conducting bar "the Hall bar" through which a current is passed by a control circuit. Electrodes are placed on opposite sides of the Hall bar to measure the transverse voltage generated by the magnetic field. In the Hall geometry there are three perpendicular directions:

- magnetic field direction, z;
- the current direction x (J is the current density along x);
- the Hall electric field direction, y.

The Hall electric field E in the y-direction comes from the basic deflection of a moving charge in a magnetic field. In this device, the Hall field produces a voltage (proportional to the magnetic field to be measured) which is monitored, and the piezodrive is used to keep the Hall voltage constant.[5]

The sensitivity of the Hall probe device has been greatly improved using modern two-dimensional electron-gas conductors based upon the GaAs–GaAlAs system,[6] in which extremely high carrier mobilities μ can be obtained. The result is that the Hall voltage, inversely proportional to n/m^3 in the Hall bar, and also the current density J, can be greatly enlarged, transforming the utility of this device.[5,6]

The scanning tip in the scanning Hall microscope is at a corner of a chip in the GaAs–GaAlAs system.

The Hall bar is approximately 1 μm wide, which determines the spatial resolution of the device, and the sensitivity is quite high because of the small number of electrons in the buried two-dimensional electron metal. The resolution will be influenced also by the spacing of the Hall bar from the sample, typically 1 μm. This device has lower spatial resolution than the STM and also lower resolution than in a magnetic force microscope where the tip may carry a tiny single-domain ferromagnet whose size in theory and in practice can be in the 100-nm range. The advantage of the scanning Hall microscope is that its response is reliably linear with respect to the magnetic field that is being observed, a feature that is less available in the MFM. The MFM lacks a definitive calibration that would depend upon calibrating precisely the nanometer-scale domain particle at the end of the tip.

The scanning Hall probe microscope can be compared with another magnetic field mapping device, the scanning SQUID microscope (SSM). The SQUID *superconducting quantum interference detector* detects magnetic flux, the product of magnetic field times the area of a pickup loop. The sensitivity of detection of flux is high: ~0.01 φ_o ($\varphi_o = 2.07 \times 10^{-15}$ T m^2). This device has high sensitivity for local magnetic field but relatively low spatial resolution, because the pickup coil at minimum has a diameter of about 10 μm. This SSM device has been important in the study of high-temperature superconductors in which fundamental questions have been answered by SSM determination that certain loops of high-T_c superconductors, interrupted by three Josephson junctions somewhat magically trap 0.5 flux quantum inside the loop.[7]

9.3
SNOM: Scanning Near-Field Optical Microscope

A scanning microscope device based on optics is the scanning near-field optical microscope (SNOM) which could be used to map the locations of quantum dots that might be decorating nuclei of biological cells in medical research. The scanning near-field optical microscope might be described most simply as an optical fiber whose end is scanned across a surface with the intention to map light sources on the surface. Optical fiber is a broadband communications medium, in terms of quantum energy, it transmits light of about 1 eV. The single-mode optical fiber as we mentioned is a cylinder of high index of refraction glass encased in a larger coaxial glass cylinder of lower index, and the whole thing is coated with a polymer to protect it and to keep its outer surface clean. The dimensions of the two coaxial glass cylinders of the optical fiber can be adjusted in conjunction with their refractive indices so only one mode propagates along the center conductor, a single-mode fiber. The action of the higher index is to pull the intensity of the light inside the inner conductor, so the light intensity dies off exponentially as one goes out into the lower index fiber. The intensity of light is thus small at the outer polymer interface. Acting as a receiver, this open-ended fiber will pick up light propagating vertically from the surface, aligned along the axis of the fiber, whose photon energy is greater than 1 eV, thus whose wavelength is shorter than about 1.24 μm. This is operated as a scanning microscope, but probably scanned at constant height to produce an image. The spatial resolution would be determined by the dimensions of the central fiber, but can be smaller than the wavelength. Another tunneling phenomenon, the near-field or evanescent electric field will allow spatially resolved reception of light into the fiber from closely spaced sources on the surface, but the resolution cannot be better than the dimension of the inner glass diameter.

In this chapter we have described the scanning tunneling microscope, imaging and manipulating single atoms and molecules. We mentioned briefly that the STM can be used in a spectroscopic way to measure the vibrational frequencies of single molecules adsorbed to a surface. We have talked about the atomic force microscope, based on a tip carried by a spring or cantilever. We described, in particular, the way the van der Waals attractive force is used in the attractive mode of operation of the AFM. A technological offshoot of the atomic force microscope is the proposed high-density nanoelectromechanical data storage device called *the "Millipede,"* which we will return to in Chapter 11. We have also mentioned scanning devices to measure magnetic field and light intensity.

10
Magnetic Resonance Imaging (MRI): Nanophysics of Spin ½

MRI *magnetic resonance imaging* is widely recognized to be at the forefront of medical diagnostic technology.

10.1
Imaging the Protons in Water: Proton Spin ½, a Two-Level System

This technology is based on the spin 1/2 and magnetic moment[1] μ_N of the proton. The proton carries a magnetic field as if it was a tiny bar magnet, and this nanophysical property is the basis for MRI. On this basis we can call MRI a part of nanotechnology, the engineering use of (sub)nanoscale elements and properties. The imaging process maps the density of protons in the specimen. In a biological sample, this translates into the density of water in the specimen. Water (H_2O), contains two hydrogen atoms, and each hydrogen atom, as we have said, is a proton accompanied by an electron. For the purpose of our discussion of MRI the magnetic moment of the proton is the focus.

The MRI apparatus appears as a large cylindrical opening where the patient is located. This large cylinder is a solenoid, a coil of superconducting wire, which provides, inside the opening, a strong magnetic field, about 1 T. The strong magnetic field is parallel to the axis of the cylinder and is uniform over the interior where the sample (patient) is placed.

How does this apparatus work? We know that the result is a digital picture presented as a grayscale or color-scale image, a cross section perhaps of your knee. The image represents the density of water in that portion of your body. In the traditional MRI image, white means absence of water, and, in practice, that will mean bone. Blood, which is mostly water, will show up as dark in such an image. In modern medical practice there are modifications and

Understanding the Nanotechnology Revolution, First Edition. Edward L. Wolf, Manasa Medikonda.
© 2012 Wiley-VCH Verlag GmbH & Co. KGaA. Published 2012 by Wiley-VCH Verlag GmbH & Co. KGaA.

enhancements, but the basic idea is to map the locations of water. The spatial resolution is at least 1 mm, which we will adopt in our discussion.

So the device can easily detect the amount of water in a one mm cube, $10^{-9}\,m^3$, which we can call the pixel, or image-generating element.

10.2
Magnetic Moments in a Milligram of Water: Polarization and Detection

So how does this work? Let us return to the protons that are detected. The proton has a magnetic moment μ_N and it has spin $S = 1/2$. This means that this moment can have only two orientations in a magnetic field B, up or down. The effect of the magnetic field on a magnetic moment[1] is to lower the energy if the moment is parallel to the field and raise the energy if the moment is opposite to the field. The difference in the two energies in the physicist's notation is $\Delta E = g\mu_N B$, g for the proton is 2.79, and μ_N is the "nuclear magneton." This is $\mu_N = \mu_B/1836 = 9.27 \times 10^{-24}\,J/T/1836 = 5.0 \times 10^{-27}\,J/T$, which is the Bohr magneton for the electron, divided by 1836, the ratio of the proton to electron mass. B is typically 1 T in the MRI apparatus, 10 000 Oersteds, about 10 000 times the earth's magnetic field. This ΔE is tiny, we have said that the proton is perhaps the smallest possible magnet. So it is remarkable that the resulting small magnetic dipole fields can be measured, as the basis for MRI.

If we pursue this discussion from the quantum mechanics, the energy difference $\Delta E = g\mu_N B$ corresponds to a frequency, through the relation $E = hf$. This is the Larmor frequency, denoted f_o, in the presence of the basic magnetic field B_o. For the proton f_o is 42.7 MHz for $B_o = 1\,T$, conditions close to those in medical magnetic resonance imaging.

The detection is done by *nuclear magnetic resonance*. Detecting the proton magnetic resonance, a difficult achievement, was first done by Edward Purcell, who won the Nobel Prize in 1952. It is difficult to detect this resonance, especially for a proton rather than an electron because the strength of the interactions for a given field is smaller by the ratio of the masses, which is 1836. One can only detect the difference in numbers or populations between spin up and spin down. The only detectable quantity is that difference called the spin polarization or alignment fraction P. The polarization P of the proton magnetic moments in the water is small because the energy difference $\Delta E = g\mu_N B$ is less than the thermal energy $k_B T$, and in nature the chance that one will have an excited state occupied versus a ground state is given by the Boltzmann factor $\exp(-\Delta E/k_B T)$, where k_B is Boltzmann's constant and T is the temperature. The MRI apparatus works at the temperature of a human being, which we will take to be 300 K. This gives, for $B = 1\,\text{T}$, a ratio $(\Delta E/k_B T) = 2.78 \times 5 \times 10^{-27}/1.38 \times 10^{-23}\ 300 = 3.39 \times 10^{-6} = 2x = 3.4\,\text{ppm}$. The polarization P has the formula $\tanh(x)$,² but $\tanh(x) \approx x$ because x is small. The result is that the polarization P, the fraction aligned along the field is about 1.7×10^{-6}, about 1.7 ppm. This is the fraction $P \sim 1.7\,\text{ppm}$ of the moments that are aligned along B, which enables us to estimate how many effective polarized magnetic moments we need to detect.

It is easy to estimate how many protons this will be, since the density of water is $1\,\text{g cm}^{-3} = 1000\,\text{kg/m}^3$, and each water molecule has two protons and weighs $18 \times 1836 \times 9.1 \times 10^{-31}\,\text{kg}$.

So in a mm cube = $10^{-9}\,\text{m}^3$ we have $2 \times 1000 \times 10^{-9}/(18 \times 1836 \times 9.1 \times 10^{-31}) = 6.6 \times 10^{19}$ magnetic moments. The polarization $P = 1.7 \times 10^{-6}$ reduces this to 1.13×10^{14} polarized spins per pixel. This may seem like a large number but it is not, as we will see.

10.3
Larmor Precession, Level Splitting at 1 T

How does the magnetic resonance detection work? Let's take a classical point of view and then we will come back and relate that to the quantum point of view. A magnetic moment has an angular momentum, like a top, and a magnetic field will exert a torque, a twisting of magnetic moments in the direction perpendicular to the magnetic field B_0. The precession phenomenon is familiar from precession of a spinning top. In the case of the magnetic moment the torque is again such as to cause the magnetic moment to precess around the magnetic field, B, direction, the z-direction. This precession is at the Larmor frequency, $f_o = 42.7\,\text{MHz}$ for a proton in 1 T.

We can calculate f_o from quantum mechanics because we know from the de Broglie relationship that $hf_o = g\mu_N B$ and so $f_o = g\mu_N B/h$, where h is Planck's constant. We can also calculate it from a classical point of view and we get the same result. The Larmor frequency f_o is a constant times B, denoted γB in the physics literature. So here $\gamma = 42.7\,\text{MHz/T}$.

How can we detect the magnetic moments per pixel precessing at 42.7 MHz, to make the MRI image? The scheme for detection can be described quickly and then in more detail, but remember that several Nobel Prizes were awarded to the distinguished people who first figured out and demonstrated these effects.

The scheme for detection is to take the polarized proton magnetic moment, which we have said is aligned along the B_o direction, and to tip it down so that it is at 90 degrees, or $\pi/2$ radians from the vertical direction. This orients the moment in the x, y plane where it will precess around B_o at 42.7 MHz.

If we can achieve this condition for all of these polarized spins, then we can detect these spins because they will produce a magnetic field also rotating at 42.7 MHz. A magnetic moment has a magnetic field pointing out along its axis, like that of a bar magnet. The strength of the dipole magnetic field is $B(r) = 2 \times 10^{-7} \mu_{pixel}/r^3$ (in T) where r is the distance from the center of the magnetic dipole μ_{pixel} arising from the spins in the (1 mm)^3 pixel.

For 1-mm spatial resolution, the imaged "pixel" mass is about 1 mg and thus contains about 1.13×10^{14} net nuclear moments, NP, aligned along an assumed 1 T magnetic field. If we think of these net nuclear spins in one pixel as forming a magnetic dipole (like a bar magnet), its magnetic field on the axis at a distance of 0.5 m is about 2.5×10^{-18} T. This rotating magnetic field induces a voltage in a pickup coil according to Faraday's law, $V = -d\Phi_M/dt$.

If we assume a 100-turn, (0.1 m)^2 pickup coil at 0.5-m spacing from the pixel (in the direction perpendicular to the polarizing field B) would provide a voltage, $V = 0.108$ nV at 42.7 MHz. A tuned detector operating at 42.7 MHz measures this signal. This is the desired indication of the water density in the pixel of interest and is fed to an element of the final grayscale image.

Now let's return to the nanophysics question, which is to how to persuade the magnetic moments to turn from their vertical orientation into the horizontal orientation, so that they precess around the magnetic field that allows them to be detected.

10.4
Magnetic Resonance and Rabi Frequency

To understand how the reorientation is done, let's imagine that we are living right with the precessing moment. In physics we would say we are going to observe from a rotating frame of reference, one that rotates exactly as the moment rotates. So we're hanging on to this moment and we're precessing with it at 42.7 MHz around the magnetic field. So our frame of reference in this thought experiment is rotating, but once we're in that frame everything is quiet. The moment is stationary in this frame of reference. But we want to flip the moment from vertical to horizontal in this frame of reference. *So we supply a horizontal stationary magnetic field B_1 in this frame of reference.* Now, the magnetic moment will precess around B_1 and the corresponding frequency or tipping rate in the rotating frame is $f_1 = \gamma B_1$ where γ is the same value that gave us 42.7 MHz for 1 T. Here, B_1 is the amplitude of an applied magnetic field whose frequency is f_o (in the lab frame). This precession will rotate the spin from up to down, the desired reorientation.

So the method of tipping the moment is to apply an alternating magnetic field of strength B_1 at frequency $f_o = \gamma B_o$, long enough so that at its precessional frequency f_1, it tips just 90 degrees. This time is called the time for a $\pi/2$ pulse, $t_{\pi/2}$, and the magic condition is $f_1 t_{\pi/2} = \frac{1}{4}$, since 90° is ¼ of a period 360° of rotation.[3] The magic frequency is called the Rabi frequency and this scheme was invented by Isidor Rabi, a physicist at Columbia University, who won the Nobel Prize in Physics in 1944. So at every pixel in the MRI measurement, an oscillating magnetic field B_1 is first applied for a time $t_{\pi/2}$ that is just enough to start the moments spinning around the static magnetic field B_o so that the pickup coil can detect their number through the strength of the voltage induced by the Faraday effect.

Building up the image requires moving the point of observation through the volume of the sample. Briefly, the "pixel" location of observation is scanned through the sample by small adjustments in the local static B magnetic field values. Magnetic resonance imaging is similar to the scanning electron microscope (SEM) in that the scanning function selects the region of signal origin (emitted electrons in the case of the SEM) while the detector is fixed.

To review, we have seen that in conventional MRI, the ac voltage from the pickup coil scales as the *square* of the polarizing field, B, because its strength contributes linearly to the polarization, P (the effective number of moments), and also linearly in the Faraday detection. This feature of the detection strongly favors use of larger fields leading to more expensive instruments. There is a trend toward 3 T values of B in state-of-the-art MRI machines, increasing the resolution or speed of measurement at higher equipment cost. Conventional MRI equipment sales in the United States were estimated at about $2 billion in 2008.

To review, we have explained the role of nanophysics in the detection aspect of the MRI apparatus. This depends on the peculiar property of $S = 1/2$, that a given single spin can be put into a state with equal probability of spin up and spin down. In this state it rotates around the z-direction with the frequency $\gamma B_o = g\mu_N B_o/h$, where h is Planck's constant.

10.5
Schrodinger's Cat Realized in Proton Spins

From the quantum-mechanical point of view, the precessing state is an equal linear combination of the spin up and spin down states. We will never know for a single spin which way it's pointing unless we measure it. If we measure a single spin we will find either up or down and if we keep doing it we will get 50% up and 50% down, heads or tails, half the time it will be up and half the time it will be down. When it's in its coherent precessing state, it's equally in the up state and the down state. The friends of E. Schrodinger, mentioned earlier, might recognize this as a "Schrodinger cat state."[4]

The imaging and data handling aspects of magnetic resonance imaging are also great accomplishments. (A Chemistry Nobel Prize in 1991 was awarded to Ernst for development of pulse sequences, such as those producing spin-echoes, beyond the $\pi/2$ pulse.) The imaging scheme is basically to select which location is sampled by producing over the sample a slightly varying or tilting magnetic field strength. The detection scheme requires that the local value of the magnetic field, $B_o \sim 1\,\mathrm{T}$, has to fulfill the condition $hf_o = g\mu_N B_o$, We assume that f_o is fixed and $B_o(xyzt)$ is slightly changed in time, to move the precise resonance location (the observed pixel) around the sample. If we change slightly the magnetic field, only those protons that experience exactly the right field will rotate and put a voltage in the pickup coil. So the scheme is to make the field strength $B_o(xyzt)$ slightly change in 3D so that in a given time only one pixel matches the condition for the magnetic resonance. This is not hard in principle but in execution it's an accomplishment. Paul Lauterbur shared a Nobel Prize in 2003 for these developments. There are small linear gradients in the strength of the magnetic field in the x-, y-, and z-directions and there are changes applied repeatedly to scan this magic location through the volume of the sample.

10.6
Superconductivity as a Detection Scheme for Magnetic Resonance Imaging

Now we want to say a few things about this and suggest an alternative scheme for making a cheaper MRI apparatus. The salient features of the existing MRI apparatus is that it requires a large magnetic field, at least one Tesla, in order to have high resolution in the image, and also to have a big enough signal from each pixel to be measured. The number of unpaired moments is extremely small, controlled by the small difference in the Boltzmann factors for spin up and spin down, $\exp(-\Delta E/k_B T) = \exp(-g\mu B/k_B T)$, which means that the polarization $P \approx g\mu B/2k_B T$, is very small and increases proportional to B. A second reason in the conventional MRI system for large B is in the Faraday's law of detection that depends upon the rate of change of the induced flux in the pick-up coil. Since a higher frequency (higher B) thus means a higher induced ac voltage. So this is a second feature favoring large B. In sum, the advantage goes as the *square* of B, within the conventional MRI method. The designers and manufacturers and doctors are always of course seeking best performance and best performance within this technology favors magnetic fields that are larger. The default value is one Tesla but we think that a more common value is 1.5 T, and state-of-the-art instruments that are close to 3.5 T. These are expensive instruments.

The weak point of the conventional technology is the lack of sensitivity in detection of the induced magnetic field coming from the precession of the moments in the one milligram pixel of the sample. If the sensitivity of the magnetic-field detection could be improved, a better and more economical apparatus might result.

10.7
Quantized Magnetic Flux in Closed Superconducting Loops

Our next topic is an improved detection scheme for magnetic field that could be used in MRI. This has been demonstrated in a simplified, cheaper form of MRI apparatus. This cheaper MRI detection scheme goes back to the nanophysics of a small loop of wire. The application will be to a small loop of superconducting wire, but the discussion can go back to the very simplest current loop, which is the orbit of an electron around a proton. An electron orbiting around a proton is a current loop in the simplest sense and we came across two ways in our discussion of the hydrogen atom of deciding what allowed orbits would be possible. First, Niels Bohr discovered the rule $L = n\hbar$ for the angular momentum around the loop. We said later that this condition was the same as having an integer number of de Broglie wavelengths, i.e., electron wavelengths $\lambda = h/p$, around the loop. Another way of saying the same thing is that the phase of the wavefunction that describes the orbiting motion of the electron in one loop has to advance by 2π per loop. The wavefunction in the angular coordinate φ, the motion around the x, y plane has to be periodic in 2π. If the electron goes around once or twice or three times what you come back to has to be exactly the same state, so the wavefunction in that angle coordinate has to be periodic in angle with period 2π.

A similar condition applies, if we have a superconducting fluid of electrons in a metal wire that form a loop. The new condition is a bit more detailed, although the condition still is that the phase of the wavefunction around the loop has to be a multiple of 2π. But in this superconducting case the phase derives from the interaction of the electron with the magnetic field, and the result is that only those currents around the loop enclosing the magnetic field are allowed that provide an *integer number of flux quanta*. This gives

$$\Phi_o = h/2e = 2.05 \times 10^{-15} \text{ W} = 2.05 \times 10^{-15} \text{ T m}^2 \text{ through the loop.}$$

Now this flux quantum is a very, very small amount of magnetic flux, 2.05×10^{-7} Oe cm^2, and was difficult to measure. It was found that in a loop of superconducting wire in a magnetic field, that measured flux through the loop jumps in integer units, exactly as predicted, verifiying $\Phi_o = h/2e = 2.05 \times 10^{-15}$ W. There is no question this works.

This nanophysical quantum-mechanical phenomenon applies only when there is a single wavefunction governing the motion of the carriers, which usually means an atomic-scale system. A coherent motion of electrons is needed, which is not true for electrons in a copper wire. The required coherence occurs in the *superconducting state* of a metal like lead or niobium or niobium tin. All of the electrons in a superconductor behave in exactly the same way and the quantum mechanics then applies even to a huge number of electrons, as given by Avogadro's number 6×10^{23} particles per gram mole.

Physicist John Bardeen was quoted that "the phase is coherent over miles of dirty lead wire." He meant the phase of superconducting pairs and he meant superconducting lead wire. He and his students, Cooper and Schrieffer, who shared that Nobel Prize in Physics, discovered the origin of superconductivity from a theoretical point of view. John Bardeen was a modest, pleasant and helpful man.[5] He actually won two Nobel Prizes, the first was for the transistor, with Brattain and Shockley. He played golf well and among those who knew him well, it was thought that he was almost as proud of his "hole in one" as he was of his two Nobel prizes.

10.8
SQUID Detector of Magnetic Field Strength

Now what does this have to do with an improved MRI apparatus? This idea of flux quantization, which is the basis for the *superconducting quantum interference detector (SQUID)*, allows an extremely sensitive detection of magnetic flux. We've already mentioned briefly the scanning superconducting quantum interference detector microscope, which is a scanning detector of local magnetic field. What we are talking about now is the same phenomenon, the superconducting quantum interference detector. A typical superconducting quantum interference detector loop includes two Josephson junctions, which are capacitor-like devices that allow tunneling and in fact allow tunneling of electron pairs. The current through a parallel connection of two Josephson junctions, enclosing an area *A*, oscillates as a function of magnetic flux intercepted in the enclosed area A. The measured current will go through one oscillation per flux quantum intercepted in the area enclosed. This allows detection of flux to about 0.01 of a flux quantum and the strength of magnetic field that can be detected then depends upon the area of the loop, and this can be multiplied by how many turns of wire enclose the area A. This device represents the most sensitive scheme that is available for measuring magnetic field.

This SQUID detector can be used in place of the pickup coil with Faraday Law detection in an MRI machine. In this case, of SQUID detection, we are measuring the field itself, not its rate of change. For this reason there is no advantage in large magnetic fields to get a large rate of change of flux in the Faraday detector. The *superconducting quantum interference detector* detection systems can be made with a frequency response at least up to the kHz range. A SQUID-based MRI operates well with a greatly reduced B field. This has the benefit that the large and expensive solenoid to produce the one-Tesla magnetic field is not required, a much lower magnetic field will suffice.

A SQUID-Based MRI Has Been Demonstrated

Spatial resolution of the order of 1 mm has been demonstrated using detection of the proton magnetic resonance by a dc *superconducting quantum interference detector*. The proton resonant frequency was 5.6 kHz, and the static B field was 132 µT. A larger field of 100–400 mT was used to initially polarize the nuclei. The pictures are shown in Figure 10.1.

A picture of a slice of a pepper is shown Figure 10.1a, in comparison with an image taken on the whole pepper, sliced electronically, typical for an MRI image! This technology may allow widespread economical medical measurements in emerging nations or even on the battlefield.

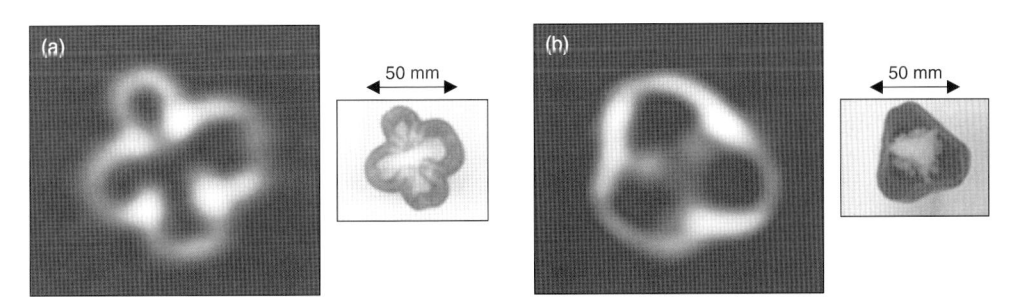

Figure 10.1 SQUID-based magnetic resonance imaging of a slice of a red pepper (a) and a similar image taken on a whole pepper, here (b) the slice is selected by the imaging technique. R. Kleiner, D. Koelle, F. Ludwig and J. Clarke, Proc. IEEE 92, 1534 (2004).

To summarize, we have explained how the magnetic resonance imaging apparatus works. The device maps the location of protons in the specimen, and in practice this means a map of the location of water molecules. Bone has very little water, and shows up as white in the usual grayscale of MRI, where black indicates a high density of water. The fraction of the protons that can be detected is given by the ratio $g\mu B/2k_B T$, which is typically 1×10^{-6} at 1 T and 300 K, to permit imaging a 1-mg pixel, which contains $\sim 1 \times 10^{14}$ magnetic moments whose precession frequency is 42.7 MHz. These are detected by being tilted into the x, y plane by a 90° pulse at 42 MHz. The rotating magnetic field is detected by induced voltage in a pickup coil based on the Faraday effect.

A more sensitive detection scheme based on the superconducting quantum interference detector, called a SQUID, is mentioned, showing an MRI image obtained in this way. This scheme allows a smaller magnetic field, but still requires cryogenic temperatures, needed here for the superconductivity of the detector unit. (Cryogenics is needed in conventional MRI only to produce the 1 T magnetic field.) The expert engineering application of the physics of spin-1/2 particles of inherent radius ~1 fm suggests that either form of MRI can be regarded as a form of "nanotechnology."

11
Nanophysics and Nanotechnology of High-Density Data Storage

11.1
Approaches to Terabyte Memory: Mechanical and Magnetic

There has been great advance in cost-effective miniature data storage, as mentioned in Chapter 2. We will discuss here two approaches, first the nanoelectromechanical "Millipede" [1] device of IBM, and secondly, perpendicular domain magnetic disk storage with magnetic tunnel junction (MTJ) readers (Figure 2.1). These competing technologies have a common size scale for the bit or pixel of storage, less than 100 nm. This bit size implies $(100 \times 10^{-9})^{-2} = 10^{14}$ bits/m^2 = 100 Tb/m^2 = 0.065×10^{12} bits/in^2. Densities of 0.1–0.2 Tb/in^2 are obtained with the "Millipede" [1].

Each device is at the state-of-the-art, and each is challenging to fabricate. The "Millipede," which has been described [1] as an AFM (atomic force microscope) cantilever array, contains $1024 = 32 \times 32$ independent cantilevers fabricated from a single-crystal chip of silicon. (A "Millipede" in biology is a segmented worm with two pairs of feet per segment, up to a total of 750.) The technical challenge in the magnetic tunnel junction approach includes that of making tunnel junctions of ferromagnetic iron with single-crystal epitaxial tunnel barriers of magnesia, MgO.

11.2
The Nanoelectromechanical "Millipede" Cantilever Array and Its Fabrication

In the "Millipede" of IBM, information is encoded as nanoscale dimples, indentations in a thin layer of polymer. The indentations are made, and then read, by tips that are precisely positioned and moved slowly across the array of data. A single tip can produce a dimple in the polymer by raising the tip temperature ~200–400°C, and lightly forcing the tip into the polymer. The tip will then be withdrawn and can be moved a small distance, approximately 100 nm, to write another dimple. The spacing of the dimples in this technology is demonstrated to be less than 100 nm. This affords ~0.1 Tb/in^2 data density that is equal to or may be better than the 1-TB magnetic disk, typically bigger than one square inch. The actual data density in the "Millipede" device probably is the highest that has been achieved.

The whole set of 1024 tips was micromachined from a single crystal of silicon, containing a buried oxide layer[1] using the photolithographic silicon technology.

How did IBM workers make a 32 × 32 array of 1024 Si tips that are free to deflect, starting from a single crystal of silicon, containing a buried oxide layer ("BOX")? This is based upon photolithography. A pattern of light is focused on a silicon surface that has been covered with a photoresist layer. A focused light image weakens a positive photoresist in illuminated portions. (Alternatively, for a negative photoresist, the light strengthens the resist layer.) The weakened portions of the photoresist are removed chemically, leaving patterned areas of bare Si. (Other portions of the silicon surface are still protected.) If we then put this silicon into a chemical etch (or etch it by means of ion beams in a vacuum chamber), we can etch the desired pattern into the exposed silicon surface.

Suppose we make a rectangular array of "mesas" of dimensions about 60 μm × 100 μm × 5 μm. (A mesa is a flat-topped hill rising above a desert floor, common in the southwest of the United States). Each mesa will resemble a 60 μm × 100 μm "shoebox" perhaps 5 μm tall. Each mesa will subsequently be patterned to include a freely deflecting cantilever.

This could be done, for example, by covering the whole silicon surface with negative photoresist, and focusing a light pattern illuminating the 60 μm × 100 μm areas, which hardens the resist polymer in those areas. Then, using a chemical bath, we remove the nonexposed photoresist, leaving the rectangular array or areas protected by hardened photoresist. A chemical bath then *etches* (removes) the silicon down to a controlled depth, leaving the array of mesa regions intact. The height is about 5 μm, including at the bottom ~1 μm of SiO_2, the BOX buried oxide layer, which will be chemically dissolved later to release the cantilever. So we have etched from the silicon surface, a 32 × 32 array of mesas of a "shoebox" shape. A triangular cantilever with an integral tip at its end will be formed along the "major axis" of each "shoebox."

The cantilevers used in "Millipede" are approximately triangular, using two support arms 50 μm × 10 μm × 0.5 μm, and a heater platform 5 μm × 10 μm × 0.5 μm holding the tip at the apex. The silicon tip of height 1.7 μm points upward at the apex. The further fabrication steps of the

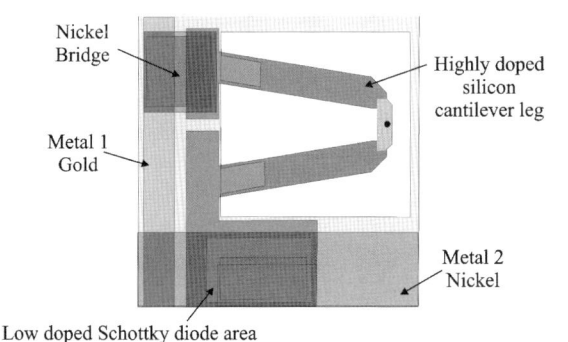

11.2 The Nanoelectromechanical "Millipede" Cantilever Array and Its Fabrication

Figure 11.1 Top view of single cell of "Millipede," about 60 μm × 100 μm, each starting from a "mesa" or "shoebox" delineated atop the Si surface by photolithography and etching (1024 such cells are etched into a single piece of silicon, which has a buried oxide layer BOX.) The data-writing tip is the dark point in the center right, just below the "highly doped silicon cantilever leg." These triangular arms form a cantilever with spring constant K about 1 nN/nm. Electrical contact to the tip's heater/thermometer "heater platform" is aided by making the slanted cantilever legs highly conducting, using ion implantation of donor impurities. These features are generally above the BOX layer that was etched away to free the cantilever from the underlying silicon crystal (Ref. [1], Figure 5).

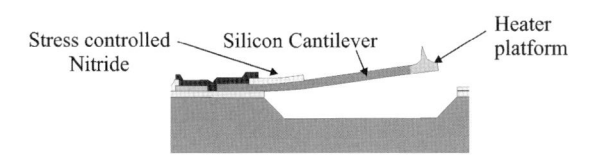

Figure 11.2 Side view of single cell of "Millipede." The white layer under the cantilever at the left is the buried oxide layer, which has been removed chemically on the right to release the silicon spring. The structure started as a Si chip with a buried oxide layer BOX, the additions shown are a "stress-controlled nitride" to induce the spring to bend upwards, and a metal contact layer on the upper left to bring current to the arms of the cantilever leading to the tip heater/tip thermometer (heater platform). (Ref. [1], Figure 5)

"shoebox"-shaped mesa into the triangular cantilever are similar. This structure is all single-crystal silicon standing on the buried SiO_2. The remarkable step that releases each cantilever so it can oscillate up and down is done by an undercutting etch that dissolves the SiO_2 under the triangular portion of the cantilever, leaving a solid support of quartz SiO_2 at the end away from the tip, to anchor it to the Si base. The thickness of the cantilever arms, as shown in Figures 11.1 and 11.2, is 0.5 μm and still leaves a section that will form the 1.7-μm tip pointing up at the other end. The spring constant of the cantilever is said to be 1 N/m and the resonant frequency is said to be 200 kHz [1].

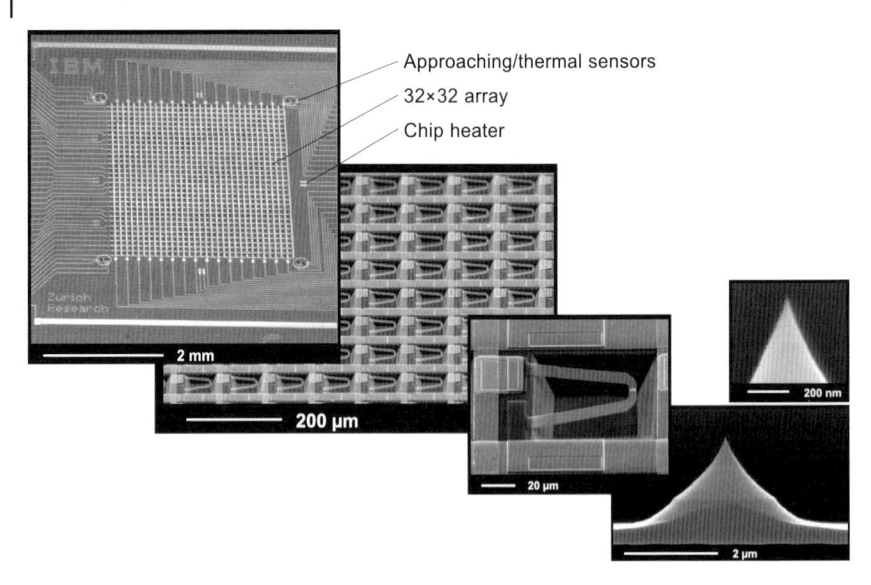

Figure 11.3 Aspects of the IBM "Millipede" data storage device, patterned and etched from a single crystal of silicon with a buried oxide layer (Ref. [1], Figure 7).

The further processing, apart from the wiring, is to pattern and sharpen the tip. Each of the 1024 identical tips, as shown in Figure 11.3, is shaped by ion beam etching, of the remaining portion of the Si above the platform near the apex. Thus, formation of the tip was done by ion-beam etching followed by a step called oxidation sharpening. It was documented (see Figure 11.3) that the radius of the remaining tip was less than 20 nm. And this could be done reproducibly to form 1024 tips on a single piece of silicon.

The wiring of this set of cantilevers included ion implantations to make the cantilever arms highly conductive to send current to the heater platform (dimensions about $10\,\mu m \times 5\,\mu m \times 0.5\,\mu m$), across the end of the cantilever. This platform region was not ion implanted, and thus was resistive, serving as a heater and as a thermometer, under the tip. The resistance of the tip heater, subsequently measured, allows to it serve as a thermometer.

The final aspect of fabrication is that the cantilever was engineered to bend upwards slightly to be the 1st element to intercept the recording polymer-coated surface, as the array of cantilevers was raised upwards towards the recording surface. The cantilever arms were bent up slightly by depositing Si_3N_4 at the join between the cantilever arms and the underlying single crystal support. After deposition it contracts slightly to pull the tip up away from the initial flat surface in the "shoebox."

So we have sketched the engineering of IBM to produce an array of 1024 independent cantilevers, located in a ~3 mm × 3 mm area. These are independently sensed but they are not independently deflected. All of these cantilevers are moved at the same time, but each cantilever has a separate heater and temperature sensor.

The operation of the system is to bring the cantilever chip precisely toward an identical silicon surface, coated with a 50-nm-thick layer of polymer, the recording area. A translation system in x, y, and z moves the two, 14 mm × 7 mm silicon chips precisely with respect to each other. The vertical motion brings the cantilevers (simultaneously) into contact with the polymer. Precise horizontal displacements of the cantilever chip with respect to the recording layer, the polymer-coated chip, are controlled to resolution of 10 nm, allowing reproducible arrays of dimples to be written spaced by 100 nm.

The means of writing from the array is to bring the tip array into contact with the polymer recording layer and to provide heating only on those tips where a dimple is to be produced. *So the whole array of tips is in contact with the polymer but only those tips desired to write a dimple are heated.* The nN-scale spring force indents the tip into the polymer to a depth of perhaps 40 nm, but only when the local tip is heated. This writing process is continued in steps by moving the cantilever array chip in units of 100 nm or so across the surface of the polymer-coated crystal.

How does this system read the information? The x-y control of the system (one chip relative to the other) is so precise as to reliably bring the array of tips exactly back above each set of written dimples. In the reading step, the tip–polymer spacing is larger than in the writing step, but, as the tips are brought closer to the dimples, the tips are gently heated and their temperature is monitored as they approach the surface, to distinguish the case of a dimple versus no dimple! Those tips that face the surface directly (no dimple) on their gentle heating come to a slightly lower temperature and thus a lower resistance. So reading the written data is achieved by reading the temperatures of the array of tips as they are brought back nearly into contact with the data set. The process also includes a scheme for erasing dimples, which we won't go into.[2]

This system is not in production. The authors [1], however, suggest that it may eventually be adopted, when the competing magnetic memories become so small that the underlying ferromagnetic domains are no longer stable. The "Millipede" faces no limit of this sort, and might evolve into a memory based on single molecules, potentially more dense than either "Millipede" or magnetic memory.

11.3
The Magnetic Hard Disk

The hard-disk magnetic disk is the leading form of high-density information storage at present. The information in the cloud computing centers is stored on magnetic disks using the nanotechnology introduced in Chapter 2, and we will now describe this in more detail. Quite differently from the "Millipede" device, now the information is written into the disk in the form of cylindrical vertical magnetic domains (tiny bar magnets) so that the magnetic field at the pixel locations, which are densely spaced on the disk surface are either up or down (see Figure 2.1). The pixel locations are arranged in linear tracks, and the detection function is based upon the relative orientation of the magnetizations in the magnetic tunnel junction sandwich. Hard-disk devices are *based on* nanophysics, as well as having nanoscale dimensions. (In comparison, the cantilevers of the "Millipede" are based on classical physics but are implemented in nanotechnology.)

First, the reading device (Figure 2.1) is a magnetic tunnel junction (MTJ), a completely quantum device. Secondly, the information is stored in the spin directions of electrons that again is a nanophysical effect. The spin of electrons comes into the operation of the disk drive in two ways. First, the *ferromagnetic domains* are regions of metal that have decided to have all of the electron spins point the same way. Secondly, the reading step, tunneling of an electron from one side to the other is a spin-specific process, which is allowed for one spin direction but not for the other.

To understand the nanotechnology here, we are going to start by explaining the nanophysics that leads to the storage medium, the ferromagnetic metal film, which appears both in the magnetic tunnel junctions and in the bit written into the disk. To understand ferromagnetism, we are going to go back to basics, in the form of basic interaction between two electron spins. The two electron spins align parallel in a ferromagnet, but align antiparallel in the hydrogen molecule. The energy involved is electrostatic (therefore large), based on the Coulomb interaction, $U = k_C e^2/r$ that we first saw in Bohr's model of the hydrogen atom.

Understanding ferromagnetism is the basis of hard disk-storage (this topic[3] is more technical, and might be omitted in a first reading!)

To understand ferromagnetism, it helps to first understand the 2-electron covalent bond, as present in a *hydrogen molecule* H_2, in itself worth understanding.[3] In Chapter 6 we introduced H_2^+, a one-electron system, which represents the simplest covalent bond. We found that the symmetric linear combination of wavefunctions for an electron on the left and an electron on the right, $\Psi_S = 2^{-1/2} (\psi_a + \psi_b)$ gave a binding energy of 2.65 eV for H_2^+ because the electron had an opportunity at the center of the structure to sample both proton attractions at once. This is a "Schrodinger's cat state" of great utility.

The hydrogen molecule, with 2 electrons, is the next step in our understanding of nanotechnology. $\psi(x_1, x_2)$, *a two-particle wavefunction*, is useful to discuss two electrons with parallel spins, as in a ferromagnet or two electrons with antiparallel spins as in H_2.

$\psi(x_1, x_2)$ describes one particle at x_1 and the other particle at x_2, as in H_2 with two electrons. The probability distribution is $P(x_1, x_2) = \psi^*(x_1, x_2) \psi(x_1, x_2) = |\psi(x_1, x_2)|^2$. Because electrons are absolutely the same here and everywhere if we were to switch x_1 and x_2 there could be no change in the probability distribution. This means that $|\psi(x_1, x_2)|^2 = |\psi(x_2, x_1)|^2$. To achieve this, the exchange can either leave the sign the same or change the sign of the wavefunction. So we have two cases

$\psi(x_2, x_1) = \psi(x_1, x_2)$ (symmetric) or

$\psi(x_2, x_1) = -\psi(x_1, x_2)$ (antisymmetric).

These two cases are distinct, which divides particles in our Universe into two kinds. This importance may be surprising! Particles like electrons are antisymmetric under exchange and are called fermions. Particles like photons are called bosons and the wavefunctions are symmetric under exchange.

But so far we have not mentioned the spin, which we certainly need for understanding a ferromagnet.[2]

Thus, we have to go one more step and to remember that the electron to be completely described has a space wavefunction ψ_A or ψ_B but it also has a spin that can be up or down. So in considering the correct two-electron wavefunctions, which have to give a minus one when we exchange the coordinates (x_1, x_2) to (x_2, x_1), we also have to consider the spins. Electrons have spin ½ . The spins can combine in different ways: First, they can be parallel, and in this case the total spin is 1, and it can point, up, zero or down, so there are three states, and this is called the *spin triplet*. This spin state is symmetric under exchange of the two particles! Second, the spins can be *antiparallel*, spins add up to zero, and so this is called the *singlet*. In this case the exchange gives a minus one, so the singlet is *antisymmetric*.

Next, the complete wavefunction for the two-electron system is the product of the space wavefunction and the spin wavefunction, and this product is the quantity that has to be symmetric or antisymmetric.

We see now there are two choices! If the space part is symmetric, then the spin part has to be antisymmetric (singlet). On the other hand, if the spin part is symmetric, the space part is antisymmetric.

We can see the answer, that *in H_2 the bonding state is symmetric in space*, implying the spins are antiparallel, by looking at Figure 6.4, but imagining that each starting wavefunction now has two electrons, which means that their spins are antiparallel. (Only one electron per completely described state implies that the second electron on each side has to have opposite spin.) Then, as before, the symmetric combination Eq. (6.6) of these two (now two-electron) wavefunctions is of lower energy because the electron probability density is nonzero at the center, where electrons can sample attraction from both wells at once.

This is how the hydrogen molecule bonds, and this is a prototype for all two-electron bonds, including the hybrid directed bonds in Figure 6.5. The covalent bond is prevalent in organic chemistry, as well as in covalently bonded molecules, N_2, O_2, and H_2.

Most important for our present discussion, the symmetry-based rules have the possibility of predicting ferromagnetism. We see that the symmetry of the space function controls whether the spins are parallel (ferromagnetic $S = 1$) or antiparallel ($S = 0$). Although in H_2 the favorable space function is symmetric, this is not true in more complicated cases, and the same argument[4] then gives *ferromagnetic alignment* of spins. It happens first in Ni and Fe, two elements that can be used as electrodes in the magnetic tunnel junction MTJ devices. In these metals, every atom wants its neighbor to have a parallel spin. And if we apply that rule, we end up with all the spins being parallel. Now this interaction, going back to the original hydrogen picture is *electrostatic* in origin, because the symmetric wavefunction allows more Coulomb attraction. Electrostatic energy is the basis for the ferromagnetic alignment of electron spins, and this makes the effect strong, on the scale of electronvolts.

We have explained how two atoms of Fe or Ni might find binding for spins parallel, leading to ferromagnetism. We extend this idea to a *ferromagnetic metal*, a large number of Fe or Ni atoms with free electrons to conduct electricity. What happens is that electrons arising from the "d-states," that is states with angular momentum $l = 2$, in these metals form separate conduction bands for spin-up and spin-down electrons. The spin-up and spin-down bands are shifted in energy by the electrostatic interaction we have been describing.

The result is that in a single magnetic domain, at the highest filled energy, the Fermi level, spins of only one orientation may be present, so that there are no levels available for spins of the opposite orientation. (In other cases, the densities at E_F may both be nonzero, but different, which will still allow the device operation.) If such a metal is made an electrode in a tunnel junction, it will be able to accept electrons of only one spin orientation, and not the other. A small strip of ferromagnet such as is used in the magnetic tunnel junction (see Figure 2.1) will have a single magnetic domain, and the *magnetization* will lie along (or opposite to) the direction of the strip. Its direction can be reversed (by an external magnetic field) in the case of a "soft" ferromagnet such as NiFe, but not in a "hard" ferromagnet such as Co.

To repeat, in a ferromagnet we have two shifted bands, spin up and spin down, a large shift of the same size and origin as the difference of bonding and antibonding energies in a molecule like hydrogen, on the scale of an electronvolt. The Fermi energy may fall in the upper end of this distribution. If the Fermi energy is in one of the bands but above the top of the other band, electron spins at the Fermi energy will be fully polarized (up or down depending upon the domain orientation). If the domain magnetization is reversed, then there will be a reversal of the spin direction at the Fermi energy.

This is the situation that occurs in the electrodes of the three-layer magnetic tunnel junction disk reader (labeled "reader" in Figure 2.1). To review, a tunnel junction is like a capacitor, with two metal electrodes separated by an insulator. The circuit behavior is that of a resistor, because electrons at the Fermi level on one side can tunnel to empty electron states on the other side. In the magnetic tunnel junction, the energy bands for spin-up and spin-down electrons in each electrode are shifted by the ferromagnetic exchange interaction (see Figure 11.4). This is the same J (of

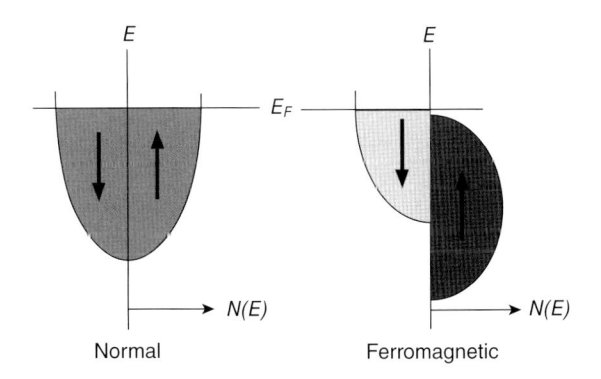

Figure 11.4 Bands for spin-up and spin-down electrons in a ferromagnet are shifted by the exchange interaction, closely related to the covalent bond energy. This may (right panel) leave only spin-down electrons at the Fermi energy, and a system that cannot accept a spin-up electron, leaving the possibility of switchable high resistance in a spin-polarized tunnel junction. This picture applies to the ferromagnetic elements shown in Figure 2.1, including the thin-film electrodes of the "reader," the magnetic tunnel junction ("Magnetoelectronics", Gary. A. Prinz, Sciencemag: 10.1126/science.282.5394.1660).

electrostatic origin) discussed for the hydrogen molecule, but it has the opposite sign.

The device is conductive (ON) when the electrode magnetizations are parallel. If we shift the Fermi energy of one with respect to the other by a small voltage V, then we get tunneling of spin-up electrons from one electrode to the other, an Ohmic response, like a resistor. This is the on condition, a low resistance. On the other hand, if we flip the magnetization: by applying a reverse magnetic field to the structure, the soft magnetic film will flip its magnetization, and now in that electrode *there will no electron states of spin up at the Fermi energy* (see Figure 11.4). Now the device is OFF. There will be no current for small bias voltage because the tunneling process preserves spin orientation, and there will be no correct final states, no current, and the resistance is infinite.

So we have almost completely understood the magnetic tunnel junction device, which is used the most dense disk drives capable of 1 TB information in one unit at the $100 cost level. The magnetic tunnel junction is carried in the reading head of the disk drive (marked "reader" in Figure 2.1), which floats above the surface of the spinning magnetic disk. The engineering of this is advanced to maintain the spacing between the reading head and the surface on the nanometer scale. The information is written into the disk in the form of cylindrical vertical magnetic domains (tiny bar magnets) so that the magnetic field at the pixel locations, which are densely spaced on the disk surface are either up or down. The pixel locations are still arranged in linear tracks, and the detection function is based upon the relative orientation of the magnetizations in the magnetic tunnel junction sandwich. The MTJ electrode films are oriented vertically (marked "reader" in upper left in Figure 2.1) and perpendicular to the track, so that the magnetic field coming up from the surface is parallel to the two electrodes, and the magnetization of one of the electrodes is fixed (as in a hardware store magnet), but the other one film has a soft ferromagnet like "Permalloy" (a Ni Fe alloy) and the localized magnetic field at the read head location above the surface domain will reverse the orientation of its magnetization. This layer is called the free layer. As the disk spins, the magnetic orientation of the free layer responds to the local magnetic field and modulates the resistance of the MTJ over a large range, approaching a factor of 2 in resistance. The tunnel barrier might be 1 nm or less. The overall MTJ device (Figure 2.1) appears of the order of 30 nm in thickness. So the tunnel junction successfully picks up the signal from one domain whose lateral dimensions are of the order of 100 nm × 100 nm, but extend more deeply into the disk surface.

The magnetic disk memory has been described based on tiny ferromagnetic domains in the disk and a magnetic tunnel junction reading device.

These devices are used in disk drives of laptop computers and of server computers in cloud computing. These devices are doubly based on nanophysics, as well as having nanoscale dimensions. First, the reading device is a tunnel junction that is a completely quantum device, and the information is stored in the spin directions of electrons, which again is a nanophysical effect.

In this chapter we have talked about two approaches to high-density memory both capable of storage of 1 terabyte, 8×10^{12} bits in a device on the scale of inches. The first was the nanoelectromechanical "Millipede" device [1] of IBM, which can be described as an array of AFM (atomic force microscope) cantilevers. The second topic was how the electron spin plays a large role in the leading source of dense memory, the hard-disk drive with magnetic tunnel junction reader. Along the way we learned that antisymmetry against electron exchange is a rule of nature that affects the covalent bond in the hydrogen molecule as well as the ferromagnetic alignment of electron spins in metals like Fe and Ni.

Reference

1 Vettiger, P., *et al.* (2002) The "Millipede" – nanotechnology entering data storage. *IEEE Trans. Nanotechnol.*, **1**, 39.

12
Single-Electron Transistors and Molecular Electronics

12.1
What Could Possibly Replace the FET at the "End of Moore's Law"?

In this chapter we describe two candidates to replace the field effect transistor FET in a post-Moore's law nanoelectronics. These are the "single-electron transistor" and molecular electronics, which includes the use of carbon nanotubes. We mention two methods of using carbon nanotubes for memory. A purely molecular logic and storage device based on 16 benzene rings is mentioned. Other candidates for replacement of "CMOS" (the conventional silicon logic family), and that also represent major changes in the organization of a computer, are described. These are an advanced superconducting logic system, and the possibility of making a quantum computer using qubits or quantum bits in place of the binary bit of conventional computing. These will be covered in Chapter 13.

12.2
The Single-Electron Transistor (SET)

Formally, a single-electron transistor (SET) is similar to a field effect transistor (FET), in that both have source and drain contacts and a gate electrode. However, the SET, schematically sketched in Figure 12.1, represents the extreme limit of a single electron controlling the device. The condition for the operation of the device is that a capacitive island, playing the role of the channel in the FET, is so small that the electrostatic charging energy[1] for single-electron charges on the island exceeds the thermal energy $k_B T$. Passage of charge through the island is controlled by modulating its potential by a gate electrode. Such a device has to be physically small, generally smaller than semiconductor technology can provide on a routine basis, and

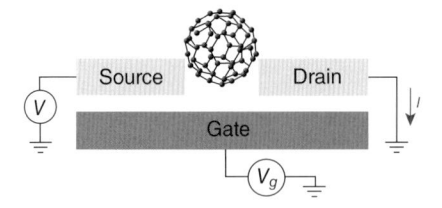

Figure 12.1 Sketch of C_{60} serving as island in a single-electron transistor. Electrical measurements showed all the features expected for SET operation, in which the current flows by tunneling of single electrons from source to drain via the island (Ref. [1], Park et al., Nature 407, 57 [2000]).

this has limited the application. On the other hand, it does not mean that the current through the device has to be small. A nA = 10^{-9} A is equivalent to one electron per 0.2 ns, a long time on an atomic scale. In a logic device context a nA is not too small a current to be useful.

The "channel" of a single-electron transistor, now called the island, is connected by two tunnel junctions in series, one at each end of the island, which plays the role of the channel, with tunnel resistance $R_{tot} = R_s + R_d$. The source and drain contacts are tunnel junctions, and the capacitance of the island is now an important parameter in the operation of the device. The source and drain tunnel junctions are assumed to be resistive, and single electrons randomly tunnel between electrode and island. A sketch of such a structure is shown in Figure 12.1.

The island, if initially uncharged, will, after random tunneling from the electrodes, be occupied by an integer number of electrons. The voltage across the whole device will be the sum of the junction voltages. The residence time on the island is simply RC, so the current, when the device is "on" takes a simple form $I = V_{sd}/R_{tot} = ne/RC$, where n is an integer. This is a staircase function with current plateaus separated in voltage by e/C.

This new behavior of interest appears when the total capacitance, $C_{tot} = C_s + C_d$ is so small that changing the charge on the "island" exceeds $k_B T$. To go from zero electron charges to one electron charge increases the electrostatic energy by $\Delta U = (e^2/2C_{tot})$, which we can call the single-electron charging energy. The capacitance is the total capacitance of the island with respect to its surroundings, including source, drain, and gate.

If this charging energy value is larger than $k_B T$, then the single-electron charging by thermal fluctuation, needed to get conduction, becomes unlikely. If the charge on the island cannot change, then no current can flow. To get $e^2/2C_{tot}$ bigger than 0.025 eV, the value of $k_B T$ at room temperature, requires C_{tot} less than[2] 3.2 aF = 3.2×10^{-18} F. The capacitance of an isolated quantum dot of 5-nm diameter, for example, is in this range.

A classical estimate of the probability of a thermal fluctuation costing energy ΔU is $P = \exp(-\Delta U/k_B T)$. If we want to make the fluctuation very unlikely, we can require $\Delta U = 10 k_B T = 0.25\,\text{eV}$, which will reduce the current to $\exp(-10) = 4.5 \times 10^{-5}$, then the island capacitance needed is $C = 3e/0.5\,\text{V} = 3 \times 1.6 \times 10^{-19}\,\text{C}/(0.5\,\text{V}) = 0.96\,\text{aF}$. The size scale of the required island again is on the scale of 5 nm, essentially impossible to construct. So, practical SET devices used in research are operated only at low temperature.

The remarkable thing about the single-electron transistor is that the average occupation of the island by half an electron turns the device on. How does this work? The role of the gate electrode, as we will see below, is to allow adjustment of n, and even to permit n to take noninteger values. The electrostatic charging energy is $Q^2/2C$, which is reduced to $e^2/2C$ in the circumstance of the island with only one extra charge on it. We are still assuming this energy even with one electron is bigger than $k_B T$. To learn how to turn the device on, we have to think of an additional fact, which is the following. In addition to having electrons jump on and off the island, it is also possible to continuously vary the total charge on the island. This is because a small electric field can slightly polarize the surface of the island to produce a continuously varying charge. This is like induction charging in electrostatics. It is a well-known phenomenon. So by this means, by varying the total charge on the island including the induced charge, the field electrode can adjust the total charge continuously. If the total island charge is adjusted to $e/2$ or actually to any odd multiple of $e/2$ then this device is turned ON.

Why is the device ON when the island charge is $e/2$? If the island has a charge $e/2$ (by careful variation of its surface charge from the gate electrode) then if we add or take away an electron, choosing the right sign, we can come to the same electrostatic energy because $(e/2)^2/2C$ is the same as $(-e/2)^2/2C$ (after adding one negative electron). So island charge $\pm e/2$ is the turn-ON condition, so, with this proper bias, an electron can hop on the island without any energy cost and that's the ON condition for the single-electron transistor. The current in the ON condition is $I = V_{sd}/R_{tot} = e/RC$. Now, this scheme gives a current versus gate voltage characteristic $I(V_G)$ that is periodic with period e in island charge. With variation of gate voltage, the device is always ON when the total charge on the island is an odd multiple of $e/2$. At even multiples of electron charge on the island, there is an electrostatic barrier at least $e^2/2C$, and since we assume this energy is bigger than $k_B T$, current is blocked.

A notable accomplishment [1] of molecular electronics is demonstration of a C_{60} molecule as the island in a SET. As we have said the single-electron transistor is a nanosized version of a field effect transistor. The characteristics of the resulting single-electron transistor were measured and followed the SET rules. In addition, this work revealed a vibrational frequency within the C_{60} molecule.

How was the single-electron transistor with a single C_{60} island fabricated?

The nanofabrication scheme (Figure 12.1) was to take an array of fine wires deposited on an insulating surface, and to artfully break these wires by passing a "large current." Consequently, a local weak spot in the wire melted, leaving an atom-scale gap between two parts of the wire. The wire-breaking process was carefully monitored, measuring the resistance R of the wire during the steps at low temperature. The wire resistance R slightly increases if a break is starting, and becomes very large, if the wire actually separates. The process involves local heating and narrowing of the wire near the location of the break. Small pulses of current were used with continuous monitoring of resistance to remove the last atom in the bridge between the two wire sections. Removing the last atom actually can lead to a large but measurable resistance characteristic of tunneling between the two, now disconnected portions of the fine wire. This leaves two electrodes within tunneling distance of each other, and the location is favorable for a C_{60}. A solvent carried buckyballs over the broken wire, which sometimes attach to the spot where the break occurred. This spot is more attractive to the C_{60}, by Van der Waals forces, than a flat surface, because the molecule will see metal on both sides instead of just in one direction. In this difficult fashion a device consisting of a single C_{60} molecule between two tunnel electrodes was constructed, putting the molecule into position as the island of the SET. Characteristics were measured [1] and followed predictions of the model of a SET.

Clearly this is a not a manufacturable process for a transistor chip! If a manufacturing method could be found, the devices would benefit from the extreme small size of C_{60}, leading to tiny capacitance, high charging energy and room-temperature operation, and also from the reproducibility inherent in a unique molecule. So, to find a manufacturing scheme for these devices is a challenge for the next generation of electrical engineers!

12.3
Single-Electron Transistor at Room Temperature Based on a Carbon Nanotube

Operation of SET devices has been achieved [2] at room temperature by engineering a single carbon nanotube. A few micrometers long nanotube was attached to two Ohmic contacts serving as source and drain electrodes. The nanotube rested on an oxidized Si chip, which served as the gate electrode. The island was formed by creating tunnel barriers, spaced by a distance about 20 nm along the center portion of the nanotube, thus the region between became the island of the SET. The tunnel junctions were created by bending or "kinking" the nanotube sharply with an AFM tip. So the device was described as a "twice-kinked" metallic nanotube. Each kink provided a ~0.25 MΩ tunnel barrier and the conducting region between the kinks formed the island. The island capacitance was so small that the SET operating condition is achieved at room temperature.

What is described [2] is a room-temperature single-electron transistor formed within an individual metallic carbon nanotube by manipulation with an AFM. The nanotube connecting source and drain gold electrodes lies on a Si/SiO_2 substrate/gate, and initially has a gate-voltage-independent resistance of 50 kΩ. The AFM tip is manipulated to produce two buckles or kinks in the tube. After the two kinks are produced, the room-temperature resistance of the tube is increased 10-fold to ~0.5 MΩ.

Measurements at 300 K of the dI/dV of the "twice-kinked" nanotube, indicate that the SET gap closes at gate voltage about −0.8 V, corresponding to half-integer electron occupancy of the isolated section of the nanotube, between the two kinks (strong tunnel barriers). The capacitance (with respect to a ground plane), and the capacitive one-electron charging energy of a 20-nm length of nanotube of diameter 1.4 nm, can be estimated using elementary physics,[3] give a charging energy $e^2/2C$ larger than $k_B T$ at 300 K (≈0.026 eV). This work demonstrates SET action at room temperature, but does not suggest a clear road to practical applications.

Reliable production of room-temperature SET devices is not possible in present silicon technology. However, useful low-temperature SET research devices have been fabricated by e-beam lithography. A version of the research single-electron transistor device with radio-frequency readout has been described, called the RFSET.[4]

12.4
Random Access Storage Based on Crossbar Arrays of Carbon Nanotubes

Carbon nanotubes of finite length can be regarded as single molecules, or as elements that can be imported into a silicon chip to form a hybrid structure. An ambitious proposed use of carbon nanotubes as random access memory RAM was described [3] in 2000 as a crossbar array of nanotubes.

The data-storage element is formed by two orthogonal carbon nanotubes, about 20 nm long, displaced vertically ~2 nm, the array is suspended above a silicon substrate. The tubes are described as single-wall "(10, 10)"[5] nanotubes, with diameter 1.36 nm. The upper tube along the *x*-direction crosses the lower tube along the *y*-direction. The spacing between the two crossing tubes is 1–2 nm. The lower tube is supported on an oxidized Si surface and the upper tube (along the *x*-axis) is supported by pillars spaced 20 nm apart, and is free to deflect. The upper tube can be bent down toward the lower tube by putting a voltage to the upper tube. Once they get closer the van der Waals force will be stronger, and the two tubes will *latch* into a closely spaced configuration. This latched condition will be retained even if the voltage is now removed. The zero and one of this storage element are the *undeflected* and the *latched* configurations of the upper horizontal carbon nanotube (Figure 12.2). The design is such that in the latched configuration the tunneling resistance between the tubes will be measured to be smaller, and this will form the readout [3].

The supports for the upper tubes are pillars 20 nm apart, so the upper tube is a doubly supported cantilever of the type we described in Chapter 3 as having a resonant frequency inversely proportional to the square of its length L. Since f is proportional to L^{-2} and L is only 20 nm, the resonant frequency of the upper nanotube is estimated to be 100 MHz [3].

A large array of nanotubes is proposed, providing nanometer intersections (junctions). The two parallel sets of nanotubes are displaced vertically, so the spacing between (upper and lower) nanotube surfaces (a design

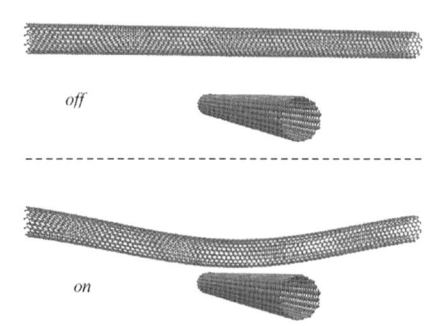

Figure 12.2 Sketches of crossing carbon nanotubes, the storage element the "Carbon Nanotube-Based Nonvolatile Random Access Memory for Molecular Computing" of Rueckes *et al.* Science 289, 94 (2000) (Ref. [3]). (Upper) Relaxed (10,10) single-wall nanotubes of nominal 20 nm length in the "off" state are clamped at each end (not shown) and are spaced 1–2.2 nm, large enough that electrical measurement shows an open circuit between the two nanotubes. (Lower) On condition (latched) is achieved by an electrical pulse pulling the upper nanotube down toward the silicon substrate. The balance between the van der Waals attraction and the energy to elastically distort the nanotube sets the spacing in the latched state. Electrical current now flows by tunneling between the two nanotubes, serving as a readout for the "on" state (Figure 2b of Ref. [3]).

parameter), ranges from 1 to 2.2 nm. The conductive nanotubes are used as control lines to apply pulses and to readout the resistances to determine the data locations [3].

The limiting performance of the *nonvolatile* random access memory RAM based on this nanotube technology proposal is stated as nearly 10^{12} devices/cm^2, capable of operation at 100 MHz. These parameters exceed those of existing technology. The latching property means that the data is retained if the power goes off.

The chip-level nanotubes in the crossbar array are arranged accurately forming a parallel array lying directly on an insulated (oxidized) ground plane surface. The upper crossing array of nanotubes are supported, midway between nanotube intersections, by an accurately located square array of support pillars.

Once an upper tube approaches closely enough to the crossing nanotube immediately beneath, it can lock into a near-neighbor (small tunneling resistance) configuration by the short-range van der Waals attraction, as discussed in Chapter 9. A subsequent electrical voltage pulse (simultaneous along the *m* and *n* control lines) can restore the upper tube to its relaxed, linear, and high-resistance configuration. The energy of the crossing two-nanotube system is given as

$$E = E_{\text{van der Waals}} + E_{\text{Elastic}} + E_{\text{Electrostatic}}$$

The van der Waals energy is negative and becomes increasingly negative as the upper tube deflects toward the lower tube.

The elastic energy can be crudely estimated (for larger separations) as

$$E_{\text{Elastic}} = (1/2)Ky^2$$

where K is the effective spring constant for deflecting the nanotube and y is the displacement of the center of the upper nanotube (toward the fixed nanotube). A simple estimate of the electrostatic energy is

$$E_{\text{Electrostatic}} = (1/2)CV^2$$

where C is the effective capacitance between the crossing nanotubes (which increases as their spacing is reduced), and V is the voltage difference between the crossing nanotubes.

The authors state that the capacitance in their geometry is on the order of 10^{-7} pF. This tiny value, 0.1 aF, is useful here in reducing the energy cost of switching and suggesting fast action.

The authors have found that attractive voltages of 4 V or more are needed to establish latching, and that repulsive voltages of 20 V or more are needed for release.

The net binding energy for 4 V attractive bias is estimated as about 4.5×10^{-18} J = 28.1 eV for (10, 10) single-wall nanotubes. This is large compared with thermal energy.

The latching, sensing and release were demonstrated on single junctions as sketched in the figure. In the latched ON state the tunneling resistance of the junction is substantially reduced, perhaps from 1 to 0.1 GΩ. The latched distortion is stable, until a second pulse is received to drive the two tubes apart.

The *fabrication* of such a revolutionary nonvolatile RAM is a formidable challenge. Nonetheless, it has been predicted that computers of the future will be constructed from three-dimensional arrays of carbon nanotubes.

Another innovative nanotube-based approach to computer memory is that of Jang et al. (2008) [4] using vertical arrays of carbon nanotubes (Figure 12.3).

The standard dynamic random access memory DRAM cell consists of a capacitor and a transistor, as the capacitor dimensions are reduced in scaling, the oxide becomes so thin as to allow tunneling and thus loss of the information. This appears to be the same problem as discussed in Chapter 8 for the FET. (That tunneling leakage was avoided by going from thermal SiO_2 gate insulator to "high-kappa" HfO_2.) Here the problem is solved differently, using a nanotube-based device.

In this case, the storage capacitor and related switch are formed by a pair of vertical multiwall carbon nanotubes.

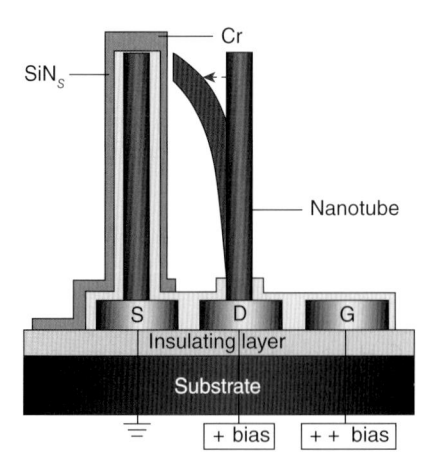

Figure 12.3 Cylindrical capacitor (left) can be charged by deflection of nanotube on the right, which constitutes an electromechanical switch. Vertical orientation reduces area of chip taken by the memory cell. Conventional memory cells on this size scale are said to be leaky (Ref. [4]. Jang et al. [2008] Nature Nanotechnology 3, 26).

By making the capacitor vertical, in the proposed scheme, the footprint of the memory cell can be reduced without encountering a leakage problem. The result is an array of pairs of vertically oriented nanotubes. These tubes are grown using a chemical vapor deposition process with Fe or similar particles to define the growth sites. In operation, a voltage pulse induces one tube to bend toward the other, to transfer charge to the second nanotube, patterned with two layers to become the storage capacitor This cylindrical storage capacitor, with overall diameter about 200 nm, is built around a carbon nanotube by depositing in succession an insulator, 40 nm of Si_3N_4, and 30 nm of Cr metal.

The array of vertically oriented pairs of carbon nanotubes is fabricated starting with 300-nm thick SiO_2 insulator on silicon substrate. Nanotubes are grown [4] on catalytically defined sites to diameter ~60 nm and controlled height. A single completed cell is shown in Figure 12.3. The capacitor on the left forms when the nanotube assembly is covered with ~40 nm of insulator (Si_3N_4), followed by metallization (Cr) of the left side of the nanotube to a thickness ~30 nm. This forms a capacitor on the left of diameter about 200 nm.

After the insulator is removed from the right nanotube, it is regarded as a flexible electrode that can bend into contact with the outer capacitor plate of the left nanotube, under the influence of positive voltages applied to the drain and gate electrodes. Contact thus charges the capacitor plate positively. After the gate voltage is removed, if the elastic restoring force exceeds the van der Waals and electrostatic forces, the switch will open.

The capacitance of this cylindrical capacitor[3] device is in the fF (10^{-15} F) range.[6]

The authors propose to add conventional silicon CMOS circuits to read the state of the memory cell. This device is by no means as difficult to fabricate as was the crossing array of nanotubes suggested in Figure 12.2.

12.5
A Molecular Computer!

Benzene rings, which are flat hexagonal C_6H_6 molecules, have been as used the basis for a 17-molecule computing assembly [5]. The carbon–carbon bond length in the ring is 0.1399 nm, so this is a tiny object. Imagine a deck of cards, one molecule being one card, the device is one card lying flat on a surface surrounded by 16 equally spaced cards on edge, surrounding the central molecule! Organic chemists are able to modify C_6H_6 in a variety of ways. Putting oxygen atoms on each end instead of H gives a benzoquinone

$C_6O_2H_4$ for example. An additional substitution is replacing the 4 H atoms by methane molecules, CH_4. So the formula for the molecule "DRQ" in this work might be written $C_6O_2(CH_4)_4$, but a more complete notation will be needed to specify at which position on the ring a given substitution is located. A self-assembling ring of such DRQ-substituted benzene rings, each a planar structure of overall device diameter about 2.5 nm, has been fabricated and studied in detail by [5] Bandyopadhyay et al. 2008. The underlying substituted benzene has oxygen atoms at each end, this is called (benzoquinone) with four methyl groups on each side. A single benzene ring is lying flat on the surface at the center, surrounded by 16 vertically oriented spectator molecules, which self-assemble to a stable configuration [2].[7]

The orientations of the methyl groups are the data storage aspect of this device. The data capacity of the assembly is considered to be $4^{16} = 4.29 \times 10^9$ as appropriate for 16 objects each of which has four distinct states. This can be considered as 32 bits, alternatively. The assembly protocol is to evaporate DRQ on a 77 K gold surface, where they lie flat and have a tendency to cluster on the basis of van der Waals attraction. The authors discovered that maneuvering an additional DRQ into the opening of a 16-fold semicircular flat cluster triggered its self-assembly to a circular array of 16 *vertically oriented* molecules encasing the last DRQ. In all cases, the oxygen substitutions are flat on the surface.

Further, they discovered that conformation changes on the central DRQ, induced by introducing charge from an STM tip, triggered domino-like changes in the conformations of the 16 spectator molecules. The fact that the paper [5] is entitled "A 16-bit parallel processing in a molecular assembly" suggests that two separate states can independently be established in the outer-ring molecules, rather than four. So the data capacity is 2^{16} in practice. For an assembly with overall diameter 2.5 nm, the data density is $16/(2.5\,\text{nm})^2 = 2.56 \times 10^{18}$ bits/m^2 = 1641 Tb/in^2, or 205 TB/in^2. This is at least 100 times larger than mentioned for the "Millipede" device!

This is certainly not a practical device, but it does make clear that the data density available in molecular storage methods is much greater than that achieved in the magnetic disk memory and the IBM "Millipede" devices.

In this chapter we have introduced the single-electron transistor (SET), logic and storage devices based on C_{60} and carbon nanotubes, and a completely molecular computing and storage element based on benzene rings. All of these are candidate devices to further miniaturize computers beyond the "Moore's law" regime. The SET is a logic device, much like a transistor, but represents the smallest possible size of such a device, since it allows only one electron at a time to pass through its channel. A low-temperature version of the SET is a proven research tool, will quickly measure electron charge and current on the single-electron level. For room-temperature operation of the SET, the necessary small size limits the fabrication. The technology to make the devices below 10 nm or so in size is not available and the same limitation applies to the miniaturization of the FET, the silicon field effect transistor. The difference is that the SET will actually function on sizes below 10 nm, while the silicon FET will not function on that size scale. One can argue that the performance of personal computers is so good and so inexpensive at present that going much further will not be so important! (Important improvements may occur, however, for large-scale supercomputers like Watson, and the cloud-computer server farms, which at present are extremely energy inefficient.) On the other hand, in the next chapter we will discuss a new computing principle, the quantum computer, which would extend computation to a wider range of problems. This would make a big change in the future of computation. We will also discuss the chance that a vastly more energy-efficient computer technology may come to a "tipping point" away from conventional silicon chips for large-scale data centers. Perhaps the most interesting question about the devices such as the SET, and the several molecular devices mentioned, is whether they could play a role in the emergence of the possible quantum computer.

References

1 Park, H., *et al.* (2000) *Nature*, **407**, 57.
2 Postma, H.W.C., *et al.* (2001) *Science*, **293**, 76.
3 Reukes, T., *et al.* (2000) *Science*, **289**, 94.
4 Jang, J., *et al.* (2008) *Nature Nanotechnol.*, **3**, 26.
5 Bandyopadhyay, A., and Acharya, S. (2008) *Proc. Natl. Acad. Sci. USA*, **105**, 3668.

13
Quantum Computers and Superconducting Computers

Computing machines started with simple devices like the abacus, in which data are represented by the positions of beads on a wire. More modern data-storage devices have notably included the punched card. Here, as in a cash register and the voting machine with a heavy mechanical lever, strictly defined numbers are represented in some durable mechanical form.

The quest for performance has resulted in change to ever-smaller entities representing numbers. The Moore's law era stores information in charge distributions in silicon and in magnetization directions in small ferromagnetic domains. The possible quantum computer, which has remarkable advantages in some areas of computing, stores information in a quantum state, such as the orientation of an electron or nuclear spin. This would be pure nanophysics! Whether such a radical change will actually occur is the first topic of this chapter.

Our second topic, the possibility of superconducting computers, was first extensively investigated by IBM with a large research and development group around 1980. This is worth consideration, because the continued ongoing search for more computer power and speed has led to alarming increases in electric power needed for computation. Superconducting computers use a tiny fraction of the power that is used by silicon computers.

One important consequence of rising device density has been the increase in power about 100 W per chip in 2002. The increasing economic impact is most noticeable for large installations, such as "data centers" or "server farms" for "cloud computing."

13.1
The Increasing Energy Costs of Silicon Computing

The energy cost of silicon computing [1] is having an important impact on large installations for rapid data handling and remote computing, so-called "cloud computing." The usage in 2006 of 150 billion kWh per year is an average power of 17.1 GW where 1 GW = 1000 MW. (This is about 3.7% of the US electric power usage in 2004, 460 GW.) On an economic basis at $0.06/kWh, the cost of this energy is about $9 billion per year. The pioneers in cloud computing have been Internet companies like Google, Yahoo, social networks like Facebook, and game services. These companies have built centralized data centers, containing up to 80 000 server computers, often located remotely at spots chosen for low electricity costs, as along the Columbia River in the States of Oregon and Washington, where there is abundant hydroelectric power. It has been reported that US data centers doubled their energy consumption in the five years to 2006, exceeding the electricity used by the country's color televisions. There is clearly a practical need for high-speed computers that do not use as much electricity.

Quantum computers might eventually fill this need. However, in the nearer term a superconducting binary computer technology is well known, this is the rapid single flux quantum computer (RFSQ) [2]. Devices and large families of circuits have been designed, fabricated and tested in this technology. It is clear that this technology has drastically reduced power requirements compared to present and future silicon technology. This technology requires cryogenic temperatures to maintain the superconducting state, so it would be useful only in large facilities. In the present large facilities device cooling and air conditioning are already major expenses. A large facility would see an advantage in providing cryogenic cooling for the computer itself, because the server units would no longer generate heat internally and the change would greatly reduce the cost of air conditioning. The RFSQ computer has to be cooled to its operating temperature, but once cold would not require further heat removal. It would be more like the magnetic resonance imaging (MRI) facility, which needs liquid helium for its operation but once cold does not require a large running refrigerator or air conditioner.

13.2
Quantum Computing

What is quantum computing? Quantum computing is based upon an information unit called the qubit. The simplest qubit is the spin ½ of an electron or proton. While a conventional binary bit takes values 0 or 1, a qubit has a wider range of values. This is most simply explained in terms of the properties of the spin ½. In quantum mechanics a wave function may be constructed of linear combinations of basic wavefunctions, so the state of a spin ½, which is one form of qubit, is a linear combination of spin up plus spin down. The general wavefunction would be $\psi = a\uparrow + b\downarrow$ down, with the requirement that $a^2 + b^2 = 1$. The binary bit would allow only $a = 1, b = 0$, or $a = 0, b = 1$.

The qubit value of "a" can be any number such that $a^2 + b^2 = 1$ which is a convenient normalization condition. Actually, in quantum mechanics "a" can be complex number, which we can think of as any point on a unit circle. The result is that the state of the qubit based on spin ½ can be represented as a vector pointing to any position on a unit sphere. So this is a large number of choices for the vector representing the state of the qubit, and the qubit carries more information than a binary bit. Why are we interested in quantum computers based on qubits as we have described?

The first reason is that solving particular important classes of problems can be much faster than is possible with existing binary systems. Physicist Lov Grover at ATT Bell laboratories proved [3] that in a common computer application, searching for an entry in a large file, a quantum computer can be much more efficient. In particular, if the file consists of N random entries, for the telephone book of all people in the United States, ~300 million, the conventional computer will have to search to find the correct entry and on the average will require 150 M separate steps. Using the Grover algorithm, based upon a quantum qubit, it is shown that the number of steps could be reduced from $N/2$ to about $\sqrt{N}$. In the example the number of sampling steps could be reduced to $(300 \times 10^6)^{1/2}$, which is 17 300. The number of steps will be reduced by the same factor, 17 300. So, in a search as described the *saving in steps is a factor 17 300*. In a simplified discussion this could mean that if a given facility has 17 300 servers, in quantum computing the facility could be replaced by 1 server!

A second potential advantage in quantum computing is to solve problems that *cannot be solved* in a reasonable time with conventional computers. One class of such problems is factoring large numbers.[1] For both reasons, knowledge about quantum computing is being sought aggressively.

How can a quantum computer be constructed from qubits that we can think of as individual spins? The basic approach in quantum computer is to provide coupling between qubits. A spin may be located close to a second spin, and if the coupling between the two spins can be programmed this would be the simplest form of a 2-qubit quantum computer. Interactions between arrays of qubits, providing pairwise couplings have been suggested as practical and adequate to make quantum computers up to say 32 qubits.

The physical realization of coupled qubits has been difficult. While nuclear spins present in molecules such as $HFBr_2$ have been demonstrated to provide quantum computation, even to demonstrate a form called adiabatic quantum computation, the three qubits (nuclear spins) provided in $HFBr_2$ are too few. The number of qubits in such molecules cannot be enlarged significantly. The molecular nuclear quantum computing method is not scalable to large numbers of qubits. However, these molecular systems have proven that the ideas behind this field do work.

An analog of the spin qubit is the *two-state electron system*, exemplified by H_2^+, the hydrogen molecule ion. In this system the analog of spin up is an electron on the left and the analog of spin down is an electron on the right. The symmetric and antisymmetric linear combination states we mentioned in Chapter 6 (Figure 6.4) now play the role of qubits. There are several approaches toward making practical qubits based on fabricated nanostructures that have two equivalent locations for a single electron. This is called the charge qubit. (The actual spin of the electron does not play a role in this approach.)

Another form of qubit is represented by the polarization of the electric field in an optical fiber.

A 3rd form of qubit is based on superconductivity. The nanophysics of a small superconducting loop, explained in Chapter 10, in connection with the SQUID superconducting quantum interference detector, says that the possible currents around the loop come in quantum units such that the magnetic flux through the loop is an integer number of the flux quantum, e/h. A superconducting qubit can be formed by a superconducting loop that is limited to having zero or one flux quanta. The technology of such loops is well developed in the technology of superconducting quantum interference detectors (SQUIDs) which form the most sensitive detectors of magnetic field, voltage and current that are available. The most advanced attempt to make a large quantum computer is a 32-qubit superconducting SQUID device developed by a firm D-Wave Systems [4]. Several smaller research quantum computer investigations have verified the predictions of Glover of increased speed in searching.

13.3
Charge Qubit

The simplest charge-qubit case is the hydrogen molecule ion, which was treated in Chapter 6, see Figure 6.4. The 0 and 1 states have the electron on one proton or the other, the fully entangled bonding linear combination symmetric state (equal occupation) is the ground state.[2]

If such molecule-ions are placed near each other, in principle entanglement of qubits will occur. Control and readout of such tiny objects is problematic, requiring electrode dimensions at the limits of electron-beam lithography. The latter aspects look easier if somehow the qubits can be larger in size, still retaining quantum coherence, to allow room for control wires.

The use of qubits in a quantum computer means a regular array must be set up, the qubits must be individually addressable, couplings among them must be present and adjustable. And the overriding need is that whatever linear combination state is required in the computation be coherent (not lose energy, nor jump around in its phase) over a length of time for many steps of computation.

The central problem of *coherence*, favors qubits from the class of macroscopic quantum states, easily obtained in superconductors, known as flux qubits. From our discussion of flux quantization in connection with the superconducting quantum interference detector, a persistent supercurrent supporting a single flux quantum through the loop, is a possible qubit state. Such a state should be stable in time. Excited states, analogous to the antibonding states of a charge qubit are likely to decay quickly, this is an example of decoherence.

The fact that these superconducting qubits require low temperature can be regarded as recognition of reality, that the colder the system, the less the thermal agitation, the longer the coherence (data storage) will be. Along the same vein, reacting to the central problem of maintaining coherence, a variant of the quantum computer called the adiabatic quantum computer [4], is promising. The qubit system in the adiabatic quantum computer is always kept in its lowest energy state, so that decoherence based on loss of energy cannot occur. Demonstrations of this method in small-scale systems have been carried out. Last, but certainly not least, the final states of the n qubits must be read out, to get the answer to the desired calculation. The most promising method seems to be a method called quantum nondemolition measurement. This may avoid the necessity in conventional quantum mechanics in determining for example, the a and b coefficients of a linear combination state, representing the answer, of doing repeated trials to establish probabilities leading to successive estimates of a and b with successively improved error bars.

13.4
Silicon-Based Quantum-Computer Qubits

A significant fraction of the ongoing quantum-computation work is based on silicon [5] . The phosphorus P donor in Si is a scaled hydrogen atom

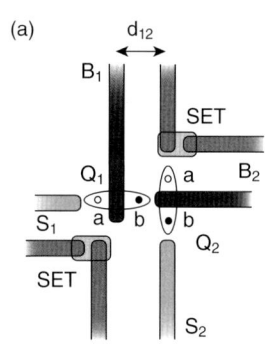

Figure 13.1 Charge-qubit coupling scheme based on the Coulomb interaction. Proposed orthogonal qubit arrangement of two single charge D_2^+ qubits, here labeled Q_1 and Q_2, each with equivalent wells a and b. Note the locations of SET readout single-electron-transistors at the "a" sites of each qubit. Note that control gates S_1 and S_2 address one well of each qubit, as well as B gates that adjust barrier heights (Ref. [6] L. Hollenberg et al., Phys. Rev. B69, 113301 [2004]. Figure 3).

(Chapter 7), and two of these can form a two-level system, a qubit, suggested in Figure 6.4. Imagine two hydrogenic donors spaced by about 20 nm, each located in pure Si perhaps 15 nm below a SiO_2 insulator, and each under a control nanoelectrode. (A sketch of such a pair of donor impurities, forming a *buried charge qubit,* is indicated in top view in Figure 13.1, on the left side, marked Q1, with sites a and b. As shown, the electron is in the b site.) If one electron is removed; then the resulting single-electron system D–D$^+$ is a charge qubit, as discussed above.[3]

Several of the aspects can be seen in Figure 13.1, first looking at the left, sites marked a and b for Q1. Q1 is a direct analog of Figure 6.4, two locations for a single electron. If the electron is started on the right (site b), as mentioned in Chapter 6, we expect that it will tunnel to the left and oscillate between sites b and a. (In the analogy to the $S = 1/2$ system, these states are the analogs of the precessing states reached by $\pi/2$ pulses, as described in connection with the MRI apparatus.) On the other hand, the stable state of the electron is in the symmetric combination $\psi_S = 2^{-1/2}(\psi_a + \psi_b)$, which is the bonding state. Alternatively, one can make state[2] $\psi_A = 2^{-1/2}(\psi_a - \psi_b)$. Starting from these states, linear combinations

$$\psi_{Qubit} = a\psi_S + b\psi_A$$

are the information storing qubits. The information is carried in the coefficients a and b, and this discussion is applicable to the single qubit Q1 on the left side of Figure 13.1. Electrodes S1 and B1 are shown that can pull the charge to one side or the other and vary the barrier height between sites a and b. Also shown is the single-electron transistor (SET) that can read out the final state of the charge, which is the final condition of the qubit.

The control electrodes $B_{1,2}$ are provided for the two interacting qubits to adjust the transparency of the tunneling barriers between the sites in Q1 and Q2. To "preinitialize" the qubit, the left electrode may pull the electron to the left while the center electrode is adjusted to prohibit tunneling out of ı L > into ı R >), which enforces the localization of the electron in ı L > . These operations are monitored by two "single-electron transistors" (SET) readout devices mentioned in Chapter 12, one sensitive to charge in the a location of Q1 and one sensitive to charge in the a location of Q2.

The "preinitialization" operation will also be the time to calibrate the SET sensors, noting the readings that correspond to one electron on the left and zero on the right. A second "initialization" step may be to provide a positive pulse of specified voltage and time duration to take the electron from the L state to the equal admixture state. *In analogy to the corresponding pulses in a magnetic spin ½ qubit,* this pulse may be referred to as a "π/2 pulse," which would mean taking the spin from vertical spin-up to horizontal, halfway toward spin down. This sets up rotation (precession) of magnetization vector around the magnetic-field direction at the Larmor frequency, generating a radio-frequency signal that is detected, for example, in the MRI imaging apparatus. In the charge qubit, free oscillations at frequency $f = h/\Delta E$ will occur after releasing the electron (increasing the barrier transparency) following the localization into ı L >. These oscillations will continue for the coherence time. After the "π/2 pulse" in the double quantum well qubit, which takes ı L > precisely into Ψ_S, the configuration will be stable until an energy loss, decoherence, occurs. Logic operations may be carried out after the several qubits involved are "initialized" in this symmetric fashion, followed by SET readout, sensitive to a small fraction of the single electron charge, $e = 1.6 \times 10^{-19}\,\text{C}$.

13.5
Adiabatic Quantum Computation

The first form of quantum computing to be proposed commercially is adiabatic quantum computing, AQC. An adiabatic process is one that proceeds with no change of energy. This computational method is based on the Schrodinger equation including the energy expression H (see Chapter 6) in its time-dependent form.[3]

In this case H represents the set of interactions that are set up between the n qubits in the system and the wavefunction is regarded as representing

Figure 13.2 A cartoon suggesting the action of adiabatic quantum computation. An initial n-dimensional state vector, known to satisfy the initial Hamiltonian, is adiabatically transformed into the nontrivial solution vector, as the Hamiltonian is slowly changed to one encoding solution of the problem at hand (Ref. [5]: Seth Lloyd, Science 319, 1209 [2008]).

information, it represents the state of each of the qubits. If the qubits are spins then the set of a, b coefficients for each of the n qubits describe the wavefunction. The time evolution of the wavefunction (information vector) can be obtained by integrating the Schrodinger equation[3] to slowly change Ψ over time. In the adiabatic case, the ground-state solution of the equation is known to adjust if the H operator is slowly changed. This approach can be argued to do computation with zero or small energy input, indicated by the word adiabatic. It also can be argued to avoid rapid decoherence, since a large class of decohering effects come from relaxation of the system to a lower-energy state. If the system is always in its ground state, such events may not occur. In practice, the energy costs of qubit rotations may not in the end be zero, but are likely to be small (Figure 13.2).

The AQC approach is to start with a known H operator with a known solution vector ψ_o, and to slowly change the operator to represent a problem to be solved. The new ground-state solution will then represent the solution to the problem of interest. The principle of adiabatic quantum computing has been demonstrated in research settings.

We now turn to a binary computing system that is quite well understood and that would clearly offer computation at lower energy costs, this is the superconducting rapid single flux quantum (RSFQ) computer technology [2]. Tunnel junctions that also exhibit the Josephson supercurrent[4] are the basic elements.

The I-V characteristic of a Josephson junction has two distinct parts. There is a current at $V = 0$ that can be increased up to the critical value, at which time a voltage abruptly appears, the value is near $2\Delta/e$ where Δ is the energy gap of the superconductor, on the scale of 1 meV, and the current jumps to a small value. (In its application to RSFQ logic the Josephson junction is placed in parallel with a resistor.) These two states, zero voltage and finite voltage on the junction, are the two binary logic states.

In a situation where a single quantum of magnetic flux, is trapped in a loop containing one or two Josephson junctions, the possibility exists that the flux quantum may leak out of the loop through one of the junctions, leading to a voltage pulse consistent with Faraday's law of induction, $V = -d\Phi_M/dt$, where Φ_M is magnetic flux. (We learned about Faraday's Law detection in Chapter 10, concerning the magnetic resonance imaging methods.)

This effect is the basis of the rapid single flux quantum (RSFQ) computer technology. The resulting voltage pulse,[5] corresponding to loss of one quantum of magnetic flux, carries the information in the RSFQ computer technology. Typically, the pulse will be a few millivolts in amplitude and last for about a picosecond.

Thus, the RSFQ approach is based on *resistively shunted Josephson junctions*, leading to a smoother $I(V)$, without an isolated spike in current at $V = 0$. The *shunting resistor* connected in parallel to the Josephson junction has a small value (perhaps 1–8 Ω) so that the current I remains nearly constant after the Josephson junction switches to its voltage state, but a transient *SFQ voltage pulse* is carried along a transmission line to the next device. The pulse is so short in time, on the scale of picoseconds, that transmitting it without loss or distortion requires a transmission line, which is fabricated on the chip.

The logic aspect is illustrated by a case where a single junction in its zero state, but biased with a supercurrent near to its critical value, may need two simultaneous incoming pulses to switch to the voltage state.

The essential test of the scheme is whether the output SFQ pulse, when presented to another biased junction, can change that junction into the voltage state. The answer is yes, in that RSFQ circuits have demonstrated counting at 750 GHz [7]. This suggests that superconducting logic is indeed rapid.

Analog to Digital Conversion (ADC) Using RSFQ Logic

One advantage of the RSFQ scheme is its fast operation, demonstrated with 750-GHz counting. A recent application is to an *analog-to-digital converter*, based on the RSFQ scheme. This device operates as a voltage-to-frequency converter, and it also involves a loop with two Josephson junctions, a so called dc-SQUID (superconducting quantum interference device). Complete ADC analog-to-digital systems with digital sampling rates ~20 GHz have been demonstrated [8]. The demonstrated ADC converters contain up to 10 000 junctions. The most immediate applications of such systems are in acquisition of radar signals. The device directly digitizes the incoming radio frequency voltage. It is anticipated that advances in closed-cycle cooling systems will lead to wider application of the superconducting devices.

13.6
Opportunity for Innovation in Large-Scale Computation

This situation suggests an opportunity for major innovation in high-performance computers, such as the servers in data centers. Most straightforward would be to replace silicon servers by somewhat higher capacity units in the superconducting RSFQ rapid single flux quantum technology. This technology is based on lossless superconducting metal, usually niobium, instead of silicon and copper, where the conventional resistance and resistive heat generation are zero. The switching of logic circuits still involves energy dissipation, but it is on a smaller scale than in silicon computers. The origin of the smaller energy scale of switching loss is that the energy comparable to the semiconductor bandgap energy, typically 1 eV, is the superconducting gap energy, typically a few meV.

We have introduced quantum computing and superconducting computing as possible approaches to satisfy the need for more energy-efficient computation, as the size of cloud computer services escalates. The energy usage for computation is several per cent of total electricity use and is rising quickly. The silicon chip is wasteful of energy requiring at least 100 W per 1 cm^2 chip, and there is no hope for a major change in this performance. Radically lower energy costs are available with superconducting computers operating in the superconducting rapid single flux quantum technology. A longer-range more profound opportunity for more efficient computing may be offered in the realm of quantum computing. In this case the data is no longer binary but the basic unit the qubit has the information contained in the linear combination states of a two-level quantum system. The advantages of possible quantum computing include much faster algorithms for some important applications such as searches, as well as solution of classes of problems including factorizing large numbers, that are essentially unavailable to binary computers. Hardware in quantum computing is at an early stage, and it is not clear that a really viable computing technology will result, even though the essential promised features of rapid search and solution of intractable problems have been verified in small-scale research studies.

References

1 The Economist (2008) Down on the server farm. The real-world implications of the rise of internet computing. Business section, Seattle, WA, May 22, 2008.
2 Likharev, K.K., and Semenov, V.K. (1999) *IEEE Trans. Appl. Supercond.*, **1**, 3.
3 Grover, L.K. (1997) *Phys. Rev. Lett.*, **79**, 325.
4 Johnson, R.C. (2007) Quantum Computer Orion Debuts. EE Times, February 8, 2007.
5 Lloyd, S. (2008) *Science*, **319**, 1209.
6 Hollenberg, L.C.L., *et al.* (2004) Charge-based quantum computing using single donors in semiconductors. *Phys. Rev.*, **B69**, 113301.
7 Chen, W., Rylakov, A.V., Patel, V., Lukens, J.E., and Likharev, K.K. (1998) *Appl. Phys. Lett.*, **73**, 2817.
8 Mukhanov, G.A., Gupta, D., Kadin, A.M., and Semenov, V.K. (2004) *Proc. IEEE*, **92**, 1564.

14
Looking into the Future

Until today revolutionary discoveries and inventions are made and are placed before us by the great minds, but a lot is yet to come. Here are some of the interesting technologies predicted by the scientists that may happen and change our lifestyle in the future.

14.1
Ideas, People, and Technologies

An interesting fact that suggests the importance of nanotechnology is that the production of *field effect transistors*, which are the central device in computers and information technology. *One billion billion* of these devices are produced per year. These devices and the computer technology are pervasive, and of course they influence our society in the tremendous increase in the flow of information. To quote from the Royal Society Report [1] the importance of the new technology is such that it "can absorb and process information on one side of the planet and deliver it almost instantaneously to the other side in a form that is immediately accessible."

One effect, now that millions of books have been digitized, is that much information is no longer hidden away in libraries, but can be gotten very quickly by using search engines such as Google and Bing. One can instantly and with no expense pull up a book written a person in the 1800s and search that book for the name of an ancestor.

These advances are revolutionizing scholarly activity. For example:

A recent study "Quantitative Analysis of Culture Using Millions of Digitized Books" [2] was based on a search of 5 million books, containing *500 billion words*. This data set was based on Google's effort to digitize books, mostly from 40 university libraries all over the world. Each page was scanned with custom equipment and the text was digitized using optical character

Understanding the Nanotechnology Revolution, First Edition. Edward L. Wolf, Manasa Medikonda.
© 2012 Wiley-VCH Verlag GmbH & Co. KGaA. Published 2012 by Wiley-VCH Verlag GmbH & Co. KGaA.

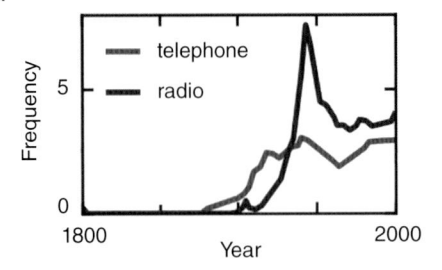

Figure 14.1 Frequency at which words "telephone" and "radio" were found in searched digitized books, containing 500 billion words. The peak in the use of the word "radio" is in 1943 [2].

recognition OCR. This data set is so large that at 200 words per minute it would take 80 years to be read. This data set can be searched quickly using Google's search engines. The study, for example, found (see Figure 14.1) that the frequency at which the word "radio" appeared in books sharply peaked around publication date 1943, one might wonder why? (Perhaps it had to do with the famous "Fireside Chats" of then US President Franklin Roosevelt, or the news toward the end of World War II that was often learned by listening to the radio?)

These advances have hugely benefited by the Moore's law growth of computing capacity. Even though the silicon chip form of Moore's law can't hold for long, new materials may replace silicon once the Moore's law collapses. Some of the promising technologies, further to those in Chapter 12, that may replace silicon transistors are molecular transistors made up of chemicals like rotaxane and benzenethiol, graphene transistors, and quantum computers (Chapter 13). High-level research is also being done on optical computers that use light beams for calculation, quantum-dot computers and DNA computers. These new technologies promise further miniaturization of computation. The size of the chip can be reduced even smaller and smaller, it can reach the size of a sand grain or even the size of a cell. This is the vision of Intel, creating shape-shifting objects using these tiny chips called "Catoms (Claytronic atoms)" [3]. These Catoms[1] have a number of tiny electrodes whose charge can be altered and are spread on their cubic structure. The principle behind is simple: these reprogrammable Catoms can repel or attract each other by changing the static electric charge on the surface. They will have memory to store or remember but the static electric charges would not be as strong as the interatomic forces and also there are a lot of bandwidth and programming issues associated. If these obstacles are overcome then a new world is going to be created. Imagine changing the size and shape of your laptops, cell phones and cars whenever you want. This might be possible in the future [3].

A lot of the advance has come from recognition and use of very basic aspects of matter. For example, "*spin*" is built into most of the particles of nature: electrons, protons, neutrons, all have spin ½. It is very simple, the spin can either be up or down, that is all there is to it. The essential *quantum* or *nanophysical* aspect of the spin, is that not only does have a state up and a state down, but it can be prepared in a state that is equal amounts of up and down. This linear combination property is extremely important. Linearity applies to quantum-mechanical particles but does not apply to large systems. So the spin, if we put it into the state 0.5 up + 0.5 down, it will actually precess or rotate around the static magnetic field, and produce a rotating magnetic field. This rotating magnetic field, detected by a pickup coil, is the basis for the MRI machine, magnetic resonance imaging, which gives a picture of your knee or your head. It is an example of the linearity property in quantum mechanics, applied to the protons in water. The joke of the early physicists was that Schrodinger's cat might be equally dead and alive, if it were put into the appropriate quantum state. The answer is, this is absurd, quantum mechanics does not apply to large systems like cats. It requires a very small particle to obey the linearity condition, and a cat is not a small particle. So this is a joke. But the *Schrodinger cat state*, if we want to use that terminology, is the state that powers the magnetic resonance imaging MRI machine. It is the state of the proton spins, namely equal amounts of spin up and spin down from water molecules that is set up in order for their detection by magnetic resonance. These heavy MRI machines are already being miniaturized and in future they can be as small as a digital camera. This tiny MRI machine will be an assemblage of a wide variety of programmable chips, sensors and DNA chips. It can scan any organ in the body and diagnose the problem within minutes with the help of a computer [3].

Nanotechnology is going to change the structure of medicine [3]. With the invention of ultrasensitive sensors and DNA chips, diseases can be detected beforehand and can be cured. DNA chips will have microscopic sites with DNA fragments embedded to detect specific gene sequences. The DNA fragments can bind to specific gene sequences and can be scanned using a laser beam. These sensors can be placed anywhere in our clothes or on the body. They will silently monitor our health everyday and give an alert signal if they detect any traces of virus before they attack us. In this way we can eliminate a lot of dangerous diseases like cancer [3]. For example, the most common gene mutation in most of the cancers is known as "p53." If we can make a sensor that detects and destroys this p53 mutation, then we can avoid the formation of tumors! The innovation of tiny chips, sensors and microelectromechanical systems will take us to another level. A related initiative is to create "smart pills." Smart pills can be directed to a certain place in our body. These smart pills can be programmed, instructed and tracked. They can be even developed to perform surgeries that eliminate the complex and traumatic procedure of cutting from outside the body. A "nanopill" is a molecule to deliver drugs to a specified target, It is a smaller version of a smart pill, being sized between 10 and 100 nm. Because of the size, these particles cannot penetrate blood cells but can move into bigger cells. Hence, these smart nanoparticles can be used to destroy cancer cells. Since cancer-cell walls have large irregular pores, the molecule can seek them, release the medicine and kill them. They can actually kill the cancer cells on the spot if laser light of their absorbing frequency is focused on them. They either heat up or vibrate, killing the cell around it. A wide range of research is being done and tested on these nanoparticles. Prototypes have already been developed by some groups [3].

The Watson computer system, using tools such as Google to collect information and also to provide rapid access to that information, is able to compete with the best human players in the TV game Jeopardy.

This is somewhat related to the idea of the Singularity, an event that might occur if and when computers are commonly better at calculating than humans. If this happens, as predicted in 2045, it might change the nature of human life. One of the primary promoters of the idea of the Singularity, Raymond Kurzweil, suggests that a situation may occur where humans become closely involved with and augmented by semiconductor computing, and the aggregate of these may become more capable than humans.

The assumption seems to be that information in one's mind may be uploaded in some way to a cloud computer. We find this hard to imagine. There is very little progress in putting electrodes into a human head to extract information, although some people think this will become possible. One can argue that there is very little progress on any means of transferring the content of an individual brain to a computer file. The New York Times recently ran a story entitled "40 Years of Analysis," the 40-year history of one individual's ongoing contacts with a psychiatrist. The information transfer, it seems, was not complete, even after 40 years of expensive and time-consuming communication.

14.2
Why the Molecular Assembler of Drexler: One Atom at a Time, Will Not Work

The advance of practical artificial intelligence is unquestioned and clearly aided by nanotechnology. If the Singularity is regarded as a positive future

14.2 Why the Molecular Assembler of Drexler: One Atom at a Time, Will Not Work

outcome, it seems a risky bet. On the other hand, there are negative outcomes predicted, which we find very unlikely.

The idea was publicized that nanotechnology might produce "gray goo," new chemical forms that might appear in a nanotechnology laboratory and self-replicate on the basis of the "molecular assembler." The fear was that gray goo might engulf the world, and this is nonsense. The answer to this is the impossibility of making a molecular assembler.

The "molecular assembler" is imagined to perform the functions of a hand or a robot arm on single atoms. The proposed device selects a particular kind of atom, grasps the atom, orients it suitably, inserts it into an atomic site in a molecule or solid being assembled atom-by-atom. It is necessary for the atom in question to be detached from the assembler arm and deposited in the proper site.

Molecular assemblers have been presented as able to assemble large machines out of atoms like carbon and silicon, on reasonable time scales. Detailed plans of nanoscale shafts and bearings have been given, with the assumption that any such structure could be built up, atom by atom, using the molecular assembler. The molecular assembler has not ever been demonstrated, and is basically impossible. No such tip-like device working on single atoms is possible.

Any such tip is too large to allow access to atomic sites in a complicated structure. A small enough tip does not have facility for orienting an atom and for adjusting its property to grasp and then to release the atom. The rate at which any such device could operate is too slow to be of use in producing gram quantities of matter on an atom-by-atom basis. The device is said to work on any atom and to build any structure.

If such a tip were possible, then in addition its action will be too slow to make appreciable amounts of material. The "Millipede" device described in Chapter 11 is intended as a memory device, but we can think of its writing function as a step of nanofabrication or molecular assembly. Let's assume its 1024 tips can be used to assemble matter atom by atom.

Ignoring all questions about the facility of a single 20-nm radius tip to do the traditional functions in mechanical assembly, let us focus on how fast the assembly could possibly be carried out.

The "Millipede" stores $10\,\text{Gb} = 10^{10}$ bits with an access time of $0.025\,\text{s}$. We can optimistically translate that as a rate 4×10^{11} steps/s, which is an overestimate. (It is an overestimate because each tip can access only 1, certainly not all, of its own $10^{10}/1032 = 10^7$ sites in 25 ms.) Suppose the goal is to fabricate a diamond structure of mass 12 grams. The molar mass of carbon is 12, so the product will contain Avogadro's number, 6.02×10^{23}, of atoms. Even at the overstated rate of 4×10^{11} steps/s, the required time will be $1.5 \times 10^{12}\,\text{s}$. *This is a very long time, about 48 000 years.*

Suppose we use more than one "Millipede" device, to work faster? To fabricate the 12-g carbon structure in 1 s would require 1.5×10^{12} "Millipede" devices working at once, again overlooking other absurdities of this idea.

Suppose we assume a much faster tip operation. For a second estimate, assume a single assembler tip the highest possible rate, about 1 GHz. *At that rate, the time for a single tip to process 12 grams of carbon would be 6×10^{14} s, 19 million years.*

These rates are so slow, using even wildly optimistic estimates on tip rates and capabilities, as to make clear that *bulk matter cannot be assembled in an atom-by-atom mode* except by huge numbers of "molecular assemblers" working in parallel. The assemblers themselves must be very small and probably have to grow in numbers by reproducing themselves, as do the cells in biology.

So, again, it will not work. If we overlook this, we can move to the next step in the argument for molecular assembly. The further assertion is made that the assemblers self-replicate. Following Smalley [4], let us assume that the "molecular assembler" can reproduce itself at the same rate of 10^9 atoms/sd. Let's further assume that the "molecular assembler" comprises only 10^9 atoms. On this assumption it will ideally take 1 s to make one copy of itself; if each copy reproduces itself in turn, then in 60 s the number of "molecular assemblers" will be 2^{60}, or about 10^{18}! (This is pure fantasy, but it is similar to the fantasy upon which the nanotechnology myth of "gray goo" rests.) If all of these huge number of assemblers subsequently work at the same rate, 10^9/s, then carbon can be assembled at a rate of 10^{27} atoms/s! For comparison, this is $10^{27}/(6.02 \times 10^{23}) = 1661$ mol/s $= 20$ kg/s for molar mass 12 g for carbon. This is a large and industrially relevant rate, but it is *impossible*, since it is based on the flawed assumption of the molecular assembler.

The idea of machine assembly of matter has been refuted by Richard E. Smalley, an authority on nanotechnology, who won the Nobel Prize in Chemistry for his work in discovering and fabricating C_{60} and carbon nanotubes. So we think the danger of gray goo is nonexistent, looking to the future as influenced by nanotechnology. Nanotechnology (as differentiated from biotechnology) is therefore not to be feared from the point of view of possibly generating self-replicating organisms. These alarming ideas came from a basically flawed concept and can be rejected as impossible.

While it is conceivable that entirely different chemical environments might allow the evolution of distinct forms of biology and life, no chemical environment will allow the imagined molecular assembler. This can be rejected simply on the properties of atoms. Atoms are the same throughout the Universe.

14.3
Man-Made Life: The Bacterium Invented by Craig Venter and Hamilton Smith

On the other hand, enhancements and manipulations of biology are well known and have already been subject to regulations. The code of life, the DNA molecule, is continually modified, in a given line of life, by *mutations*, this is the basis for evolution. Breeding has always involved capturing and preserving mutations and this process has been accelerated by modern developments. The most recent advance has been the creation of a new bacterium, *M. mycoides JCV1-syn1*, by J. Craig Venter and his coworkers.[2] The DNA for this new bacterium has 1.08 million base pairs, designed and constructed at the J. Craig Venter Institute, using the bases C, G, A, T as described in Chapter 4. The authors describe their work as "writing the software of life." They suggest that creating new life in this way can usher in a new era of science and lead to new products and applications, including advanced biofuels, clean water technology, and new vaccines and medicines.

The new bacteria reproduces as shown in Figure 14.2 and is distinct from previous bacteria, so it is an artificial form of life, although rather similar to existing forms of life. Such bacteria might well be dangerous, but there are

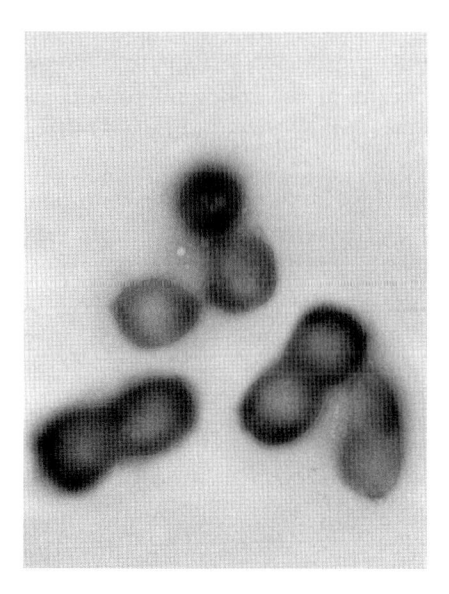

Figure 14.2 Natural cell division occurring in the synthetic bacterium *M. mycoides JCV1-syn1*. Transmission electron micrographs taken by Deerinck and Ellisman of the National Center for Microscopy at the University of California at San Diego (http://www.jcvi.org/cms/research/projects/first-self-replicating-synthetic-bacterial-cell/overview/).

plenty of dangerous bacteria and viruses already, and their propagation has been discouraged by unified action of governments and cooperation of sensible people.

The promise of nanotechnology that is quite predictable is the accelerated connectivity of human beings around the planet. Optical fibers and rapid computers have leveled the playing field for scholarship. An isolated researcher, if he has access to a computer and the Internet, can immediately learn of the latest advances in many topics and can contribute on a short time scale to the advance of human understanding. The slow print process of producing scholarly articles has been augmented by a parallel process involving journals that are available for free on the Internet. The processing speed and availability is much greater on the Internet than as published in traditional journals like The Physical Review or the Transactions of the IEEE or J. Am. Chem. Soc.

On the human level interconnectivity has been improved by social utilities like Facebook, with 800 million subscribers, where one can find one's old friends and relatives, and can be in rapid contact with them if everybody agrees. It appears that the course of human affairs has been influenced by the Internet connectivity, in particular in the revolution in Egypt. A recent article [5] entitled "The Facebook Freedom Fighter" explains how Wael Ghonim, a Google marketing executive in Egypt, after having been arrested and put in jail, used Facebook to organize protests.

The efficiency in scientific research, in communications in both business and personal spheres have benefited greatly by the computing and communications and Internet phenomena and rationally one would expect these things to continue with beneficial results.

Universal translators are already on their way on a small scale. Futuristic versions of these are being developed to give subtitles seen directly in a contact lens, or create an audio translation that is directly fed into ears, enabling us to carry on a conversation speaking in different languages [3].

We have seen above the replication of artificial life in the form of the new bacterium *M. mycoides*. Should we fear that computer systems like Watson can reproduce and take over the world?

It is easy, however, to get carried away in speaking about these things, and there are reality checks that should be applied. One form of the discussion about the suggested Singularity talks about self-replicating computers. As an experimental physicist, E.W. finds this to be very unlikely. To make a Pentium chip one needs a large, expensive plant, including machines of very rare types. Some examples are lasers that produce 193-nm light, ion implanters, and high vacuum systems working on slices of silicon. It seems unlikely that a Pentium chip can be reproduced anywhere but in such a semiconductor "fab" facility, which is a billion dollar item. Pentium chips will not reproduce on their own and a computer is always dependent on a power supply that is beyond its domain of authority. One can pull the plug on a hypothetical rogue computer and it will stop.

We have seen exponential growth in Moore's law, and this idea is the central point in the argument that there may be a "Singularity." But the earth and its resources are not infinite, as exponential growth would require.

From a different perspective, the growth of all these technologies is increasingly energy dependent, and rational people conclude that supplies of cheap energy are unlikely to continue beyond, at most, a hundred-year time scale.

Cheap energy has created what we might call a bubble, in which everything is available cheaply but much of it depends upon oil that is definitely limited in supply, and on coal that has a longer supply but poisons the air and may have to be restricted. So there is serious concern that the energy-based technology bubble may deflate when the cheap energy is not available and the standard of living may decrease for the wealthy countries. A rational response to this threat is obviously to find new sources of energy.

14.4
Future Energy Sources

Enough energy is available, but means of harvesting the energy (to create electricity or a portable fuel like hydrogen, methane, or gasoline) are lacking. Sources of energy include direct sunlight, and its secondary influences, which are winds, hydroelectric power, and the energy of water waves. The sunlight falling on an area about equivalent to the area of the existing US highways, could be converted, using solar cells, into electricity to replace entirely the power plants now in use. A rather similar situation exists with windpower. A significant fraction, 20%, of electricity in Denmark comes from wind turbines. The common problem with solar and wind energy is the intermittency of the power. This can be solved either with storage or by incorporating many locations into a common grid, with the idea that when one location is quiet another will be strongly contributing. An extensive plan was produced for continental Europe and North Africa [6], concluding that with large use of DC power transmission lines and undersea cables, which are more efficient at longer transmission distances, the whole area could be serviced with renewable energy, without need for oil and coal.

In addition, the water of the sea contains enough deuterium, if its energy is extracted by fusion reactions, to power our civilization for an essentially infinite time. The development of practical means of converting deuterium into electrical power by *fusion* of deuterons into helium and other nuclei, can in principle be viewed as a problem in nanotechnology, because it involves engineering as applied to particles on the subnanometer scale [3]. Fusion will be much cleaner than the regular uranium fission plants. But attempts to harness fusion power have been tried for a long time. There are several technical issues involved, and until they are fixed fusion power generation is not in the near future [3].

An area of increased need is clearly for energy sources to replace fossil-fuel sources. Nanotechnology is needed to develop reliable and inexpensive devices to efficiently convert sunlight to electric power. Solar cells, whose active elements are P-N junctions similar to those we discussed in Chapter 8 on injection lasers, can be made by assembling nanoparticles, called nanoinks, onto moving rolls of aluminum foil. These semiconductor devices are forms of nanotechnology with dimensions of the P-N junctions of hundreds of nanometers, and the assembly method is made inexpensive by printing an ink made of nanometer-size particles of the semiconductor. This method uses a minimum amount of the semiconductor raw materials, some of which are in short supply.

Researchers at MIT have created a practical "artificial leaf" that uses sunlight to produce hydrogen and oxygen. This is a leap in the research of clean energy. This silicon wafer has cheap catalysts coated on each side that take energy from the sunlight absorbed by silicon. These catalysts then split water into hydrogen and oxygen. This hydrogen can be used in fuel cells or to produce electricity. We can create potentially cheap and clean energy source if the artificial leaf can be developed into a full-fledged device [7].

Hydrogen fuel-cell cars, also called the cars of future, are next in line to electric cars. A fuel-cell car runs only by combining hydrogen and oxygen to produce electrical energy, leaving only water as waste. But the main problem again is the hydrogen gas that has to be created by separating water into hydrogen and oxygen by electricity, which conventionally uses fossil fuel. (Electrolysis of water at a wind farm site or by conversion via the "artificial leaf" may in future contribute.) So work is being diverted to create efficient energy. Hydrogen–oxygen fuel cells are definitely making progress, and the method is being extended to methanol and other hydrogen fuels. Some companies are working on nanotube fuel cells too. These nanomaterials are light but very strong and exhibit large surface areas for storage relevant to fuel cells [3, 8].

A new class of room-temperature superconductors are also under study, and may revolutionize a future automobile industry [3]. The superconductors have very interesting properties. They have no energy losses below their critical temperatures because the electrons all join together and move without energy loss. The great advance has been with the ceramics, with superconducting properties at 138 K, which is far above absolute zero. There is no reason why this effect might not be found even at room temperature.

Perhaps a greatly improved superconducting version of maglev (magnetic levitation used in trains in Japan and China), might be available [3]. This could greatly improve the energy efficiency of transportation, removing road friction, and using electric motors, able, when acting as brakes, to recover the energy of motion. The majority of conventional energy, like gasoline, is used up for overcoming friction either on road or in air. Possible room-temperature superconductors may change the situation, since there is no electrical resistance and no energy cost in maintaining a large magnetic field needed for levitation. But there are issues with ceramics and difficulty molding them into wires to overcome. If we solve them, we would be running into an age of magnetism. These room-temperature superconductors would be helpful in the miniaturization of MRI machines too [3].

We expect that the benign and productive areas of nanophysics and nanotechnology are likely to continue to contribute toward advance in technology. These efforts are likely to lead to further developments in areas of biotechnology, AI/robotics, renewable-energy production, such as in solar cells, as well as in communications and information technology.

14.5
Exponential Growth in Human Communication

The nanotechnology revolution is most evident in how humans can now communicate instantly across the whole world. This is recent and profound and available to almost everyone. A recent study summarized in Figure 14.3 shows *faster than exponential* growth in telecommunications capacity. The "effective telecommunication capacity" in MB in 21 years from 1986 to 2007 is shown to increase by a factor 260 from 3×10^{11} MB to about 7.8×10^{13} MB. What is not shown in the graph, but explained in the article, is that the huge rise is a transition from mostly analog (two-way) communications in 1986 to Internet-based digital communications, fixed and mobile telephones, in 2007. According to the authors, "The Internet revolution began shortly after the year 2000. In only 7 years, the introduction of broadband Internet effectively multiplied the world's telecommunication capacity by a factor of 29." In 1986 the average person could transmit/receive 0.16 MB per day (mostly via fixed-line telephone) and in 2007 the average person could

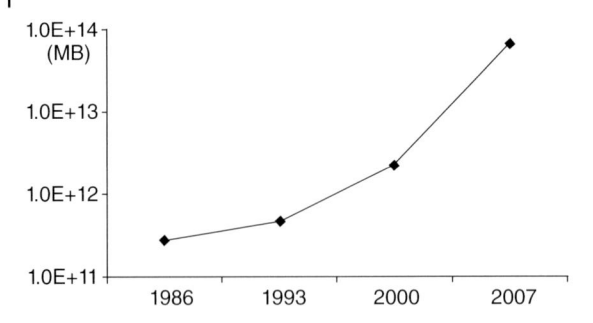

Figure 14.3 Growth of effective capacity, in MB, in analog and digital technologies of telecommunication [9]. This growth is faster than exponential! The capacity increased by a factor of 29 in the 7 years from 2000 to 2007. The peak value in this curve (see text) corresponds to about 27 MB per person per day two-way communication capacity.

telecommunicate 27 MB per day, 99.9% digital. *This is the information explosion!* The curve shows a faster-than-exponential rise, since an exponential rise would be a straight line on this plot [9].

A faster-than-exponential curve is suggestive of a singularity! But does this mean that "the Singularity," machines surpassing humans *in all areas of thinking*, is more likely?

The Singularity has already happened.

The most provocative idea associated with the growth of computing power is "the Singularity," the proposal of Raymond Kurzweil that in 2045 computers will surpass human intelligence. Further, that humans become closely involved with and augmented by computing, and the aggregate of these may become more capable than humans.

Unquestionably, practical artificial intelligence is a main benefit of the nanotechnology revolution. But this is not the same as "the Singularity." The recent event in which Watson the computer beat the leading contestants in Jeopardy emphasizes the great advance in practical artificial intelligence. But this does not really support the idea of "the Singularity," which is that computers will by 2045 *become equivalent to humans in all areas of thought* [8].

What does that mean, to be equivalent to humans in all areas? How about playing poker? Do you think that a computer ever will be able to win in a multihand poker game? The answer: "No, never will this be possible," was argued recently by Adam Gopnik [10]. Gopnik points out that the machine is rigidly programmed and is unable to make use of hints like facial expressions of other players in a poker game. As far as competing with machines like Blue Gene and Watson, Gopnik says, "we find ourselves at war . . . .

with immobile, pathetic nerds who have somehow memorized the phone book and the encyclopedia." Without denying the vast utility of artificially intelligent machines, Gopnik goes on to say "perhaps the real truth is this: the Singularity is not on its way – the Singularity happened long ago. We have been outsourcing our intelligence, and our humanity, to machines for centuries. They have long been faster, bigger, tougher, more deadly. Now they are much quicker at calculation and infinitely more adept at memory than we have ever been."

The extension of human capability by technology has been the topic of this book, with emphasis on the most recent developments in nanotechnology, particularly useful in information, computing, and human connectedness. In a real sense we can say that the Singularity, the information explosion, has already happened.

14.6
Role of Nanotechnology

As a summary, we can say that nanotechnology, as distinct from genetics, biotechnology and robotics, is a natural interdisciplinary area of research and engineering based on nanophysics, chemistry, materials science, mechanical and electrical engineering and biology, and has no particular risk factors associated with it. Nanometer-sized particles of all sorts have been part of nature and the earliest forms of industry for all of recorded history. Nanometer-sized chemical products and medicines are routinely subject to screening for safety approval along with other chemicals and medical products.

A large part of our mission has been to raise awareness and understanding of the unity among the various aspects of nanotechnology and the underlying science, which have played an important role in the several cascading technologies that have brought the present-day information revolution.

References

[1] The Royal Academy of Engineering, The Royal Society of London (2004) Nanoscience and nanotechnologies: opportunities and uncertainties: Royal Society Report 2004, http://www.nanotec.org.uk/finalReport.htm (accessed 3 December 2011).

[2] Michel, J.-B., et al. (2011) Quantitative analysis of culture using millions of digitized books.

Science, **331**, 176. doi: 10.1126/Science.1199644

3 Kaku, M. (2011) *Physics of the Future: How Science Will Shape Human Destiny and Our Daily Lives by the Year 2100*, Doubleday, New York.

4 Smalley, R.E. (2001) *Sci. Am.*, **285** (3), 76.

5 Giglio, M. (2011) The Facebook Freedom Fighter: Wael Ghonim's day job was at Google. But at night he was organizing a revolution. Newsweek, Feb. 13, 2011.

6 Czisch, G. Low cost but totally renewable electricity supply for a huge supply area – a European/trans-European example, http://www.iset.unikassel.de/abt/w3w/folien/magdeb030901/overview.html (accessed 3 December 2011).

7 Service, R.F. (2011) Artificial leaf turns sunlight into a cheap energy source. *Science*, **332**, 25.

8 Kurzweil, R. (2005) *The Singularity Is Near: When Humans Transcend Biology*, Viking Penguin, New York.

9 Hilbert, M., *et al.* (2011) The world's technological capacity to store, communicate and compute information. *Science*, **332**, 60. doi: 10.1126/Science.1200970

10 Gopnik, A. (2011) Get Smart. The New Yorker, April 4, 2011, p. 70.

Notes

Chapter 1

1 *Homo sapiens*, the earliest humans whom we will consider, originated in Africa some 200 000 years ago, and according to some accounts, coalesced into a small group surviving drought about 50 000 years ago in the "horn of Africa" from whence they migrated to populate the globe. A readable account of this is given in "The long march of everyman: It all started in Africa" The Economist, Dec. 20, 2005, with references therein. We can speculate that these early humans may have resembled the tall slender Kenyan cross-country runners who often win the Boston Marathon. By the necessity of running animals to death in a hot climate, to obtain food, they had become bare-skinned, to allow them to perspire and to run long distances without heat exhaustion. They had already learned how to make spears with light shafts and sharp points, to be thrown. This gave them an advantage, also, over a competing species, the Neanderthals, who used heavy stone axes, and had to come close to an adversary to do combat.

  The Neanderthals were apparently no match in fighting the *Homo sapiens* in being able to survive in a variety of environments. It is reasonably believed that all of today's population of 7 billion descend from small groups in Africa. A major survey of these migrations is "The Genographic Project," joint between IBM and The National Geographic Magazine. This project is aided by nanotechnology, as we will describe in Chapter 4.

2 The tools made of stone were documented from 2.6 to 1.5 million years ago in Africa and outside Africa to about 0.5 million years ago. The Bronze age dates from 3000 BC to 1200 BC and the Iron age from 1200 BC to 500 AD. The Battle of Kadesh took place between the Egyptian empire and the Hittite empire in 1274 BC. It is known to be the largest chariot battle engaging about 6000 chariots. The Hittite warriors used ankle-length mail coats made of Iron, which is a type of netted mesh armor consisting of linked metal rings for protection and the Egyptian charioteers used bronze scale mail armors as a protection in the war [3, 4, 6].

3 Astronomical methods of estimating longitude at sea were proposed, but apparently not widely used. Estimating latitude was much easier from a ship. It is clear that early navigators like Columbus used dead reckoning. One time zone (one hour's time difference) is 15° in longitude, which corresponds to 645 miles at 51.5° North Latitude (Greenwich, UK) but is 1036 miles at the equator, latitude $\varphi = 0$.

4 A nucleus is made up of Z neutrons plus N protons, giving a total nucleon number A (A is an integer telling how many "mass units," about 1.67×10^{-27} kg, are in the given nucleus). All nucleons attract each other, independent of their electrical charge: let us say U per particle. U is a large energy, about a million times larger than a chemical binding energy. So the attractive energy of A nucleons is AU, minus a small correction from the surface, proportional to $A^{-1/3}$, (i.e., $1/3\sqrt{A}$) arising simply from those nucleons near the surface without partners. On the other hand, since the rules of nuclear physics favor approximately equal numbers of protons and neutrons, $Z = N = A/2$, the charge of the nucleus, eZ, increases approximately as $eA/2$. We will discuss these topics later in Chapter 5.

It is an exercise in electricity to find that the *repulsive energy* of a uniform spherical distribution of positive charge is $U = 3/5\, k_c Q^2/r = 3/5\, k_c\, (eZ)^2/r = 3/5\, k_c\, (eA/2)^2/(R_o A^{1/3}) \sim A^{5/3}$. Since the repulsive energy increases faster ($A^{5/3}$) than the attractive energy (A), there is a point where the overall binding energy per nucleon reaches a peak value, and this is near $A = 60$. So nuclei with A larger than 60 can break up to form more stable units by a process of fission, most famously as ^{235}U ($Z = 92$) breaks up into two nuclei, each having Z around 46. The main contribution to the release of energy is the reduction in Coulomb energy as Z is reduced from 92 to two nuclei of about $Z = 46$ apiece.

5 "The past 30 years has seen a revolution in information technology (IT) that has impacted the lives of many people around the world" quoted from p. 17 of Ref. [1], the Report of the Royal Society of London. This report [1] (p. 17) states the size of the market for the IT industry (2004) as $1000 billion (about $150 for every human on the planet), and expected to rise to $3000 billion by 2020. The success of nanotechnology is such that the price of a single transistor has fallen to "less than one-thousandth of a cent" (10 nano $). The importance of the new technology is such that it "can absorb and process information on one side of the planet and deliver it almost instantaneously to the other side in a form that is immediately accessible" [1]. A recent book, "Silicon Earth," by John D. Cressler (Cambridge University Press, 2009) is subtitled "Introduction to the *Microelectronics and Nanotechnology* Revolution." (Italics added.)

Chapter 2

1 Information has no inherent size, as may be evident in our discussion. A number is an idea that can be realized or represented in many different ways. The small-

est meaningful size in nature is thought by physicists to be on the Planck size scale, 1.6×10^{-35} m. In practice, the smallest possible size of a bit might be taken as an atomic size, say a hydrogen atom, which is a sphere of radius 0.0529 nm. A binary bit could be the presence or absence of a hydrogen atom at a given location.

2 If we think of an "abacus" as an array of locations, each occupying a volume of 1cm^3 then we could say a "bit" is either having (1) or not having (0) a bead at that location. (Alternatively, at each location the bead could be up or down on a support, this would take twice the space, but is closer to a real abacus.) Then, eight locations will constitute a byte, and we will estimate how large a planar array $L \times L$ would be needed to store the 10 TB, which is conventionally the information in the Library of Congress; $10^{13} = (1/8)(100\,L)^2$ (if we express L in meters). Therefore $L = 89.44$ km, or about 55.6 miles on a side.

3 In fact, no real abacus works at all in this way, since an abacus has a set of beads that one can move about, but the total number of beads is not changed in a calculation. A binary abacus could be envisioned, if each bead has two positions, say up or down, on a supporting wire. In practice, there are many forms of ancient abacus, typically with vertical wires carrying beads, with a horizontal bar. In a typical Japanese abacus, which uses base 10, there are four beads below the bar and one above. Each lower bead is worth 1, and the top one is worth 5. If you are to add 3 and 4, you push the 3 below the bar up. Then when you get the 4, you have to push the top bead down to the bar. You find the complement of 4 in 5 (1), and so push one of the lower beads down. This abacus works completely on base 10. The Chinese abacus has five beads below and two above. Neither uses binary numbers. No early people used binary numbers. The Babylonians used base 60 (along with base 10), and the Mayans used base 20. Most others either used a tally system or base 10. Place value systems came in quite late in the game. The Babylonians had them, but the Romans did not.

4 The particular advantages of binary numbers for computers based on relays, vacuum tubes, and transistors were recognized by Prof. John Atanasoff, the inventor of the modern computer in 1939 at Iowa State College in Ames, IA, and by his contemporaries John Mauchly in Pennsylvania and Konrad Zuse in Berlin, Germany. The history of the digital computer has been explained in a recent book: "The man who invented the computer," by Jane Smiley (Doubleday, 2010). The Atanasoff–Berry computer was built in the Physics Building of Iowa State College by Physics Professor John Atanasoff and his assistant Clifford Berry. Iowa State College declined Atanasoff's request to patent his invention, and eventually junked his original apparatus, the Atanasoff–Berry computer, which was a working machine, in 1948. Thus, it happened that the most important invention of the twentieth century, the digital computer, was never patented! For when John Mauchly later filed for a patent on the computer, the court ruled that the computer as embodied in the Sperry Rand machine called "Eniac" was common knowledge (Mauchly had learned about it by visiting John Atanasoff at Iowa State College in June of 1939.)

5 Ref. Wikipedia, TSMC central Taiwan plans $9.3 billion facility for 2012, individual stepper tools in this facility may cost $50 M apiece (http://en.wikipedia.org/wiki/TSMC).

6 Iron, cobalt, and nickel are typical metallic (electrically conducting) ferromagnets, contrasted to magnetite (iron oxide, an insulator) used by the ancients, as early as 200 BC, for compasses.

7 Tunneling, or barrier penetration, by electrons through a thin insulator gives a device following Ohm's law, passing current proportional to voltage. The tunneling process is important in the MTJ because it leaves unchanged the spin of the tunneling electron, to be discussed in Chapter 5. The magnetic tunnel junction (MTJ) is a capacitor-like sandwich device whose conducting electrodes are tiny ferromagnetic metallic thin films separated by an extremely thin (2 nm) insulating tunnel barrier. The modern MTJ device has been developed as recently as 2004 through the accumulated efforts of physicists and materials scientists. The novelty of the MTJ is the sensitivity of its tunneling resistance to the relative orientation of the magnetic domain orientations in its two electrodes. The other feature is that this device can be made extremely small. Ferromagnetic electrode dimensions 100 nm × 200 nm × 10 nm are possible, tunnel barrier thickness 2 nm (W.J. Gallagher and S.S.P. Parkin, IBM J. Res. Dev. 50, 5 [2006]). The basic ferromagnetic domain, representing the stored information, is on the disk, pointing up or down relative to the surface, in the most recent versions.

8 The earliest reading device for the magnetic disk was a pickup coil that develops a voltage, according to Faraday's law, proportional to a rate of change of magnetic flux. So the magnetic field of the moments in the surface of the hard disk in the earlier case the domains in the surface were parallel to the surface in a linear fashion, like a line of bar magnets. The choice at joins is whether you have an up and a down, which would produce small or no magnetic field, or "up-up" or "down-down," which would produce, respectively, an up or down vertical magnetic field above the location on the disk. Those fields, as they were rotated past the coil of the pickup device, by their rate of change, would produce a Faraday's law voltage in the wire connected to the coil. The size of the induced voltage was amplified by use of several turns of wire wound around a soft magnetic core. Such a coil is not easily miniaturized in mass production. In summary, the pickup coil device was much inferior.

9 The prior giant magnetoresistance (GMR), or "spin-valve" devices, honored by Nobel Prizes in Physics in 2007 to Albert Fert and Peter Grunberg, was also based on nanophysics and required nanometer-size scales. The GMR device, a sandwich of nanometer-thickness copper between two ferromagnetic layers, conducts best (*longitudinally*, along the copper film) when the ferromagnetic orientations are parallel, turns off in antiparallel orientations of the ferromagnetic domains. (This aspect is similar to the MTJ device, although in the MTJ [TMR, tunneling magnetoresistance] device the current density is *transverse*, crossing the tunnel barrier layer). Accordingly, MR is defined as MR = $(R_{AP} - R_P)/R_P$. The GMR was also an example of "smaller is better," but it turns out that the MTJ

simply works better in practice, and now dominates the marketplace. TMR and GMR are both new-concept devices, which require tiny dimensions, and represent new physical mechanisms of operation discovered in the miniaturization process.

10 The central aspect of nanophysics in the great improvement in the disk drive performance, confirmed by the 2007 Nobel Prizes in Physics given in this area, has escaped notice in some quarters. For example, the Royal Society Report (Ref. 1.1) on p. 18, in its Section 3.4.3, speaks of "writing information magnetically to a spinning disk. It is therefore primarily mechanical, or more strictly electro-mechanical . . . ." The paragraph on p. 18 goes on to say "Although the individual bits of magnetic information that are written onto the disk drive . . . are currently smaller than 100 nm, the constraints related to this nanotechnology . . . require fabrication of components with even greater precision." The whole rather long piece does not include the word "physics," which, as we have discussed, has been central to the development, by providing sensor devices tiny enough and sensitive enough to read the magnetic field specific to a *single 100-nm-size magnetic domain*. The market is large, for example, the leading supplier Seagate, in the first quarter of 2010, reported sale of 50.3 million hard drive units for revenue of about $3 billion (http://www.xbitlabs.com/news/storage/display/20100421215905_Seagate_The_Hard_Drive_Market_Is_Still_Constrained_by_Supply.html).

11 The erbium-doped fiber amplifier (EDFA) provides amplification of the light wavelength near 1.5 µm in optical fiber communication. The trivalent erbium Er^{3+} ions added (doped) into the amplifier section of fiber are illuminated (pumped) by a shorter wavelength, thus excited to a state that decays quantum mechanically to add more photons (Chapter 5) to the existing light signal, thereby amplifying that signal. The pumping light signal is itself generated by a laser, a necessary part of the EDFA assembly (Mears *et al.*, Electron. Lett. 22, 159 [1986]).

Chapter 3

1 Historically, one form of spring driven oscillator (actually with rotary motions) was the hair-spring/balance-wheel combination invented by Huygens in 1675, which was early adapted to wristwatches and also was used in the Harrison chronometer. The "H4" version of Harrison (See Ref. 1.2) was 13 cm in diameter, with a total mass 1.45 kg. In a 2-month voyage at sea, from England to the Caribbean, it was accurate to 51 seconds and won the prize from the Board of Longitude of the British government.

2 This working formula (3.3) $f = 0.162 \, (t/L^2) \, (Y/\rho)^{1/2}$ is a bit messy, which is often the case in technology. (Actually the prefactor 0.162 is not easy to obtain; it requires solution of a "boundary value" problem in mechanical engineering.) Technologists have to be careful and pay attention to details. To decide if a formula makes sense, checking the units is a good start. The units of this formula

(3.3) are, from left to right, m^{-1} $(Pa\,m^3/kg)^{1/2}$. In the parenthesis we have (force $m^3/m^2\,kg)^{1/2}$ and we know the units of force (= mass times acceleration) = $kg\,m/s^2$. Collecting these we have $f = m^{-1}$ $(kg\,m^3\,m/s^2\,m^2\,kg)^{1/2}$ or $f = s^{-1}$. So the units are correct for this formula.

3 This illustrates a distinction in scaling, between 1D and 3D (isotropic) scaling, for something that is approximately one-dimensional like a cantilever whose length is L. We can scale the length L and leave the width and thickness alone, or we can scale the whole thing uniformly, which we will call 3D scaling. It turns out by taking the formula $f = (1/2\pi)\,(K^*/M^*)^{1/2}$ taking the actual form of spring constant in the elastic beam, that if we scale only the length, then the frequency goes as $1/L^2$. So if we reduce the length by factor of 10, the frequency will go up by factor 100. On the other hand, if we scale all three dimensions, the cantilever will shrink uniformly, then the frequency changes as $1/L$.

4 In practice, the quartz crystal is oriented ("AT cut") so that the motion Δt excited by the electric field is transverse to the thickness, and the appropriate modulus is the shear modulus S, whose value is about 31 GPa instead of about 100 GPa. The frequency–thickness product for AT-cut crystalline quartz is given as $f\,t = 1.67\,GHz\,\mu m$ (G. Sauerbrey, Z. Physik 155, 206 [1959]). The same value of this product is obtained using the formula (3.6) in the text with choice $Y' = S = 23\,GPa$.

5 The values plotted in Figure 3.5 are summarized here.

Tabulated values of length and frequency of different devices

	Length L (in meters)	Frequency (in hertz)	log 1/L	log f
Pendulum clock	1	0.5	0	−0.301
Tuning fork	0.066	534	1.18	2.727
Quartz oscillator	2.75×10^{-3}	32.8×10^3	2.56	4.158
Silicon nanomachined cantilever	2×10^{-6}	186.8×10^6	5.698	8.271
Quartz PC clock	5×10^{-6}	0.277×10^9	5.301	8.442
Carbon nanotube	20×10^{-9}	29×10^9	7.698	10.462
Hydrogen molecule	$0.074 \times 10^-$	1.32×10^{14}	10.13	14.12

6 *Streamline flow* occurs for values of the Reynolds number $N_{Reynolds}$ less than approximately 2000. $N_{Reynolds}$, which is dimensionless, is defined as $N_{Reynolds} = 2R\rho v/\eta$, where R is the radius, ρ is the mass density, v is the velocity, and η is the viscosity.

Chapter 4

1 The power exerted by a rotary motor is $P = \omega\tau$ where $\omega = 2\pi f$ and torque $\tau = r\,F$, where the force F is exerted at a radius r from the axle. This formula can be

reached, starting with the formula for work $W = Fx$, so that $P = dW/dt = Fv$. In the rotary case $P = \omega\tau = (2\pi f) \; r \; F = Fv$, since $2\pi f \; r = v$. The units are watts if torque is in N m (Newton meters) and ω is in radians/s.

2. The accurate value is 3.4 Angstroms = 0.34 nm, which we approximate as 0.5 nm. This is the result of X-ray crystallography that was originally used by Watson and Crick but has been enormously refined with modern use of high-energy "synchrotron light sources" to get high intensities of X-rays.

3. The binding in the "lock-and-key mechanism" is called hydrogen bonding in chemistry. The AT pairs form double hydrogen bonds, while the CG pairs form triple hydrogen bonds. No bonds are possible in the other cases, AC, AG, nor TC, TG. The details of such bonding might be considered nanotechnology treated expertly by chemists. Similar lock and key pairings are designed to make drugs that will go to specific places in an organism. These features are seen in Figure 4.2.

4. However, if the 4 km length curls into a planar spiral, like a pancake, we have $A = \pi r^2 = 4 \times 10^3 \times 2 \times 10^{-9} = 0.8 \times 10^{-5} \, m^2$. The resulting radius is $r = 1.6$ mm, which is much smaller than a TB disk drive, which is several inches in diameter. This estimate depends on the accepted value 2 nm for the diameter of the DNA strands.

5. If we allow the DNA to ball up into a sphere of radius R, so $V = (4/3) \pi R^3$, and $V = 4\,km \times (\pi/4) (2 \times 10^{-9})^2 = 1.26 \times 10^{-14} \, m^3$, then $R = 1.44 \times 10^{-5} = 14.4 \, \mu m$, for 1 TB of information. This represents extremely dense data storage, evidently $1/(1.26 \times 10^{-14} \, m^3) = 7.94 \times 10^{13} \, TB/m^3$. The average size of the nucleus in a mammal's cell is 6 μm (http://en.wikipedia.org/wiki/Cell_nucleus#Chromosomes), which gives a data storage around 8.97 GB.

6. We can make a comment about the complexity of a bacterium such as E. coli. Although the dimensions are small, perhaps 3 μm long by 600 nm in diameter, with a volume of $8.5 \times 10^{-19} \, m^3$, this is really a big system, with lots of room, in the following sense. The working units inside E. coli are molecules, really small. As an example, the energy molecule ATP with dimensions about 2 nm by 0.7 nm diameter, has a volume of only $7.7 \times 10^{-28} \, m^3$. The number of ATP molecules that would fit inside then is $8.5 \times 10^{-19}/7.7 \times 10^{-28} = 1.1 \times 10^9$, more than a billion ATP will fit inside the volume of a single E. coli. Physicist Richard Feynman made the famous remark that "there is lots of room at the bottom"; perhaps this is an instance.

Chapter 5

1. This particular bacterium is really interesting, because of its magnetism. These bacteria have flourished at the bottom of the sea for millions of years. As they die, their magnets are lined up with the direction of the earth's magnetic field, which has varied over time. The accumulating magnetic layers resulting from these dead bacteria are then a possible source for "paleomagnetism," the science

of determining the orientation of the earth's magnetic field over time. For more information, see Harrison et al., Proc. Natl. Acad. Sci. USA 99, 16556 (2002).

2. A magnetic moment experiences an energy $U = -\mu B \cos\theta$ if it is oriented at an angle θ with respect to a magnetic field B. The basic source of a magnetic moment is a current loop of area A carrying current I, the rule is $\mu = IA$, and the vector nature can be expressed by defining the direction as orthogonal to the area element A. The torque exerted on a magnetic moment is $\tau = \mu \times \mathbf{B}$, which is a vector equation. The motion of a magnet moment μ in the presence of a magnetic field is similar to motion of a spinning top that precesses around the vertical (the direction of the acceleration of gravity).

3. Notice also, from Eq. (5.1), that atoms get bigger for smaller mass and inversely. More accurately, the m in Eq. (5.1) is the reduced mass $m_{red} = mM/(m + M)$, where M is the mass of the proton. Since M/m is large, about 1835, we can overlook this detail in some cases. However, if we replace the electron with the heavier muon, mass 207 m_e, in a *muonic*, and much smaller, version of the hydrogen molecule, and if we replace the protons with deuterons, the deuteron nuclei in vibrations will get so close together to undergo fusion, observed to release energy on the scale of millions of electronvolts!

4. The hydrogen atom is used to define "atomic units," including units of time, size, and energy. Atomic timescales arising in more complicated atoms have important technological applications. *Atomic clocks* now officially define the value of the second and are essential in the operation of the global positioning system (GPS). We mentioned that the electron has a magnetic moment and also that the proton of spin ½ similarly has a magnetic moment. Magnetic moments interact with each other, as you remember from bar magnets. Magnets like to stick together if the polarities are reversed, such that a north pole touches a south pole. The same interaction is present between the magnetic moments of the electron and the proton in a hydrogen atom. The energy of the electron is slightly changed by the relative orientation of its magnetic moment with that of the proton. The basic understanding of Planck was that the energy of a photon $E = h\nu$, where h is Planck's famous constant and ν is the frequency. E in this formula can also be regarded as the *difference of two energies* in the system that is emitting or absorbing the light. If this concept is applied to the electron and proton spin moments in the hydrogen atom, the difference in energy from flipping the electron magnetic moment with respect to the proton magnetic moment emits a photon whose wavelength $\lambda = c/\nu$ is 21.3 cm. In fact astronomers use this 21.3-cm line to map the location of hydrogen clouds in outer space. The 21.3-cm line, which corresponds to a frequency $c/\lambda = 1.41$ GHz (1.41 billion oscillations per s) is specific to hydrogen. Cesium, used in the atomic clock, is similar to hydrogen in that it has a single valence electron and also a nuclear magnetic moment. In Cs the electron orbits a larger atomic core that has 55 protons surrounded by 54 electrons but still there is a spin interaction between the outer electron and the nucleus of the cesium atom, and the energy difference as one changes the spin orientation of the electron in the Cs atom with respect to its nuclear spin, cor-

Chapter 6

1. In Figure 6.2, it is the repulsive energy between the two protons forms the barrier, whose energy is $U(r) = k_c e^2/r$. If a proton approaches from the right, with $E = 1.293$ keV, corresponding to 1.5×10^7 K, the classical turning point, $r_2 = 1113$ fm $= 1113 \times 10^{-15}$ m. Since the incoming proton slows down and turns back at this radius, the wavefunction's amplitude and wavelength both become large at $r_2 = 1113$ fm. But this is far away from the spacing, 2.4 fm, needed for the particles to overlap and start the fusion process! The wavefunction $\psi(r)$ decays exponentially, as sketched in the region between $r_2 = 1113$ fm and the point of contact, r_1. At this point the two particles join to form a two-proton ^{2}He state that sometimes decays and sometimes results in a deuteron. The tunneling probability for this situation is about 10^{-8}, but requires a more detailed analysis than the square barrier. This is the first step in the proton–proton cycle that lights the sun, and each fusion event of this type, followed by other easier steps, releases 26.2 MeV. If tunneling were unavailable the energy generation of the sun would stop. This also illustrates the formation of the simplest nucleus, the deuteron is a proton and a neutron, but the initial part of the process is indeed a proton overcoming the Coulomb barrier as described. All heavy elements are believed to have been formed at high temperature in stars, following tunneling processes for which this is the simplest, which then explode as supernovas. In the words of Carl Sagan "We're made of star-stuff." Tunneling is a key process in forming the nuclei that has led to the chemical table of the elements.

2. One might think that since the temperature is so high, 15 million K, the proton might cross the barrier by "thermal activation" which often happens in chemistry and also, as we will see, accounts for the forward current in a P-N junction. The proton–proton kinetic energy is 1.293 keV, but the barrier $U(r_1)$ is 600 keV. So the probability factor $P = \exp(-\Delta \bar{E}/k_B T) = \exp(-598.7\,\text{keV}/1.293\,\text{keV}) \approx 10^{-201}$. So the tunneling process, $T \approx 10^{-8}$, is much more likely. The p–p reaction is complicated by added step of turning a transitory ^{2}He state, suggested by the well at $r = 0$ in Figure 6.2, into a D plus a positron and a neutrino. The overall probability of Deuteron formation is much less than $T \approx 10^{-8}$, but everyone agrees that the tunneling step we have described is necessary to start the reaction, which indeed accounts for most of the energy generated by the sun.

3. The uncertainty principle says $\Delta p\, \Delta x > \hbar/2$. This formula is basically a property of waves, but its most precise form, quoted here, is exactly found in the quantum treatment of the simple harmonic oscillator, the mass on a spring as was discussed in Chapter 3. Applying this to the trapped particle, Eq. (6.5), tells us that

if $\Delta x = L$, then $(\Delta p)^2/2m = (h/2)^2/(2mL^2)$, which is close to the exact answer. *The kinetic energy is increased by localizing a particle.* This is also seen in the starting point for Schrodinger's equation, $p = -i\hbar\, d\psi/dx$, where $i^2 = -1$. Localizing the particle increases $d\psi/dx$ that increases momentum, and momentum squared is energy. (The opposite effect, lowering energy by delocalizing electrons, is the origin of the metallic bond.)

4 Let's talk a bit about units and sizes. We said that the electron charge is 1.6×10^{-19} C. We can get some insight into this by thinking about a common computer accessory, the charger. The charger for my PC says it will produce 3.25 A. That is defined as 3.25 C/s. To get the number of electrons per second we divide by 1.6×10^{-19} C, so the charger will pass 2.03×10^{19} electrons/s. So the electron charge is very small, and that is why, it was so hard to observe the granularity of electron charge. The common scale of energy is a joule, related as 1 W s, and 746 W = 1 HP. The unit of energy appropriate for electrons is the electronvolt, which is 1.6×10^{-19} J, and corresponds to the energy gained by an electron in falling through a potential difference of 1 V. The electronvolt is an atomic scale of energy, the binding energy of the electron in hydrogen is 13.6 eV.

5 This is only possible for tiny systems, and is at the core of nanophysics. (This is how the world works for microscopic particles.) It does not work for large systems, for a famous example, a cat. Suppose (this does not work) ψ_A represents Schrodinger's cat *alive*, and ψ_B represents the cat *dead*, then ψ_S and ψ_A are two ways of representing the cat *simultaneously dead and alive!* This makes no sense for cats, but the idea evidently amused Schrodinger's friends. (It is said that he came up with his famous equation when on a skiing vacation, between semesters of teaching.) When linear combinations are sensible is also when quantum mechanics is useful. Such linear combinations are the basis of covalent bonds (as we describe in the text) and in *quantum computing*, where the linear combinations may be referred to as "Schrodinger cat states."

6 The normalization prefactor means that the total sum of the probability over all values of r is exactly 1. The formal description of the wavefunctions is $\psi_{nlm}(r,\theta,\varphi)$ and the relation to rectangular coordinates is $x = r\sin\theta\,\cos\varphi$, $y = r\sin\theta\,\sin\varphi$, $z = r\cos\theta$.

7 The states are $\psi_{2,1,\pm1}(r,\theta,\varphi) = c'\, r\, e^{-(r/2a)} \sin(\theta)\, \exp(\pm i\varphi)$. Adding and subtracting these for $m = \pm1$ generates $\cos\varphi$ and $\sin\varphi$, respectively. These linear combination states no longer carry angular momentum, but are pointed along the x- and y-axes, respectively.

8 The same electronic states, described by quantum numbers n, l, and m are used in describing atoms with Z electrons (matching Z protons). The Z electrons are added into the states of the spherical atom, limited to two electrons for each specific (n,l,m). For example, for carbon, $Z = 6$, we have two electrons in $n = 1$, "$1s^2$," plus four electrons in $n = 2$, namely two in (2,0,0) "$2s^2$" and two in (2,1,0) "$2p^2$."

9 The weak force plays a role in the reactions of elementary particles, but does not strongly influence the binding energy per nucleon. The weak force is the origin

of "beta decay" (electron or positron decay) of nuclei. An example is the reaction of two protons (Figure 6.2) to produce an unstable ^{2}He state that sometimes goes to a final state of a deuteron, a positive electron, and a "neutrino" v, a neutral particle of nearly zero mass. The "weakness" means that the probability of the fusion as indicated in Figure 6.2 as about 10^{-8} for the tunneling step, actually leads to the deuteron only in about 10^{-25} of the successful tunneling cases. The experts on stars say that this "weakness" greatly extends the lifetime of stars. There are so many protons in the sun that this extremely unlikely reaction p + p → D + e$^+$ + v, followed by faster steps leading to ^{4}He and 26.2 MeV, is the source of the sun's light.

10 The same rule, two particles per state, applies when we think of modeling a metal by adding electrons to an empty box with rigid walls. The trapped particle energy levels increase, as we saw in Eq. (6.5), as the square of the quantum numbers n, and only two electrons will be possible within a given completely described state.

Chapter 7

1 In this case we are using e = 2.718 is a more convenient base than 10. To get a decimal expression, we would change the formula to $P = 10^{-\Delta E/(2.303\,kT)}$.

2 The localization energy depends on the number of half waves across the crystal in the x-, y-, and z-directions, as $E_n = [h^2/8\,mL^2](n_x^2 + n_y^2 + n_z^2)$. This formula (Eq. (6.5a)) is also correct for localized hole states in semiconductor "quantum dots."

3 It has recently been discovered that carbon nanotubes are present in the famous Damascus steel blades, swords and sabres. The Damascus blade was a technological advantage employed by the warrior Saladin, who used these sabres or swords of great sharpness and flexibility. In that respect it has been learnt that these Damascus blades were made of high-carbon steel that came from India that also contained heavy metal impurities, perhaps manganese or cobalt. A recent study of steel samples from actual Damascus sabres revealed multiwalled carbon nanotubes. We have mentioned earlier that Fe is used as a catalyst for nanotube growth, Co and Mn are likely to act in a similar fashion. The recent TEM pictures show, without doubt, nested carbon nanotubes, in a sample of Damascus steel. These carbon nanotubes had six or eight nested layers and provided in transmission electron microscope, an unmistakable identification (by Fourier analysis of TEM images) of the expected spacings between layers in nested carbon nanotubes. Damascus steel technology is an example of a lost technology, similar to the lost technology of processing the wood in Stradivarius violins, another example from materials science.

4 So in a certain sense this is a superconductor, without a resistance, because of its perfection. It turns out that this statement is not quite correct, even at zero temperature. Resistance will be not really zero, however, because the atoms, although on a perfect crystal lattice, do vibrate in thermal motion. This can be estimated classically by setting $k_B T = ½\,Kx^2$. For spring constant $K = 1000\,\text{N/m}$

and 300 K, x_{rms} = 2.04 pm, which, though small is still 1.5% of a carbon–carbon bond length, 134 pm.

Chapter 8

1. $f(E)$ is defined as the probability that the electron state E is occupied with formula $f(E,T) = \{\exp[(E - E_F)/k_B T] + 1\}^{-1}$. At low temperature this function plotted vs. E is a step function, going from 1.0 at $0 < E < E_F$ to 0.0 for $E_F < E$. The width of the transition from 1 to 0 is about $2 k_B T$. For energy more than about $2 k_B T$ larger than E_F it is approximately given as $f = \exp[-(E - E_F)/k_B T]$. For energies more than about $2 k_B T$ smaller than E_F it is approximately given as $f = 1 - \exp[(E - E_F)/k_B T]$, so that the probability of finding a hole (no electron), $1 - f$, is about $\exp[(E - E_F)/k_B T]$. For example, at the valence band edge, $E = 0$, and $E_F = E_G/2$, the probability of finding a hole (a missing electron or broken bond) is about $\exp(-E_G/2 k_B T)$. For Si at room temperature this is $\exp(-0.55/0.026) = e^{-21.2} = 6.5 \times 10^{-10}$.

2. The values N_C and N_V are influenced by the effective mass, related to the flatness of the band, and also by the temperature T. The formula is $N_C = 2(2\pi m_e^* k_B T/h^2)^{3/2}$ that has units m^{-3} with a similar form for the valence band.

3. There is a slight shift because the number of states in the band is influenced by m^*. The accurate result is $E_F = E_g/2 + \tfrac{3}{4} kT \ln(m_h^*/m_e^*)$, for a pure sample where the number of holes exactly matches the number of electrons, both arising from a broken bond. There is a useful product rule that is implied by these formulas, namely $N_e N_h = N_c N_v \exp(-E_G/k_B T)$. This product, which is sometimes called N_i^2 is *independent of* E_F, meaning that it is unchanged by doping (impurity levels)!

4. In silicon with electron concentration $N_e = 10^{22}$ el/m³, for example, we can deduce the value of the Fermi function, since we know the band density of states, $N_C = 2.5 \times 10^{25}$. Thus, $N_e = 10^{22}$ el/m³ $= f \times 2.5 \times 10^{25}$, so $f = 10^{22}/2.5 \times 10^{25} = 4 \times 10^{-3}$. Using the formula $f = \exp[-(E_c - E_F)/k_B T]$, we can find E_F by taking the natural logarithm of the equation: $\ln\{\exp -(E_c - E_F)/kT = 4 \times 10^{-3}\}$ and we see that $E_c - E_F = 0.14$ eV. In standard notation, the value is $E_F = E_G - 0.14$ eV $= 0.96$ eV. So this is well above the center of the energy gap.

5. Similarly, we can get an analogy for the heavily doped P-type contact (P^{++} metallic contact), the Fermi energy will be below the valence band edge. This will provide a free system of holes, acting like a metal made up of positive carriers. In this case the $E_F = -(h^2/8m)(3N_h/\pi)^{2/3}$. As in the N-type case, the carrier concentration can be approximated by the actual dopant concentration, N_A using boron, for example. These contacts are sometimes actually metallic compounds, silicides which can be formed by deposition and heating as well as by ion implantation.

6. $W = [2\varepsilon\varepsilon_o (V_B - V)(N_D + N_A)/e(N_D N_A)]^{1/2}$ where the formula includes an applied voltage V, in the sense of reducing the width of the depletion layer with positive V. This shows that the depletion region disappears at $V = V_B$, which is not an

advisable value because the heating would likely destroy the device. In reverse bias one obtains an increasing W and the application is the varactor, a voltage-variable capacitor.

7. Si has an *indirect band gap* so that the electrons have a large momentum. For an electron to disappear, creating a photon of zero momentum, somehow the extra momentum of the electron must be taken up. To conserve momentum, a vibration called a phonon has to be simultaneously produced along with the photon of light. This makes the process much slower, so light emission does not occur in Si P-N junctions.

8. Optical fibers are waveguides for light, formed as long filaments of glass in a coaxial arrangement with higher-index core and lower-index cladding. The choice of the index of refraction keeps the light wave confined to the core, and the choice of material keeps the losses very low. These fibers span oceans and bring TV signals and more with less delay than the earlier satellite data connections. The best optical fibers are arranged so that a single mode of light propagation is confined into the core.

9. Why is that? It is because the index of refraction for this wavelength of light, near 1.5 μm, is $n = 3.3$ for GaAs, vs. 1 for air or vacuum. The reflection probability, defined as $R = (n-1)^2/(n+1)^2$, thus has value 0.286. This is a partially reflecting mirror.

Chapter 9

1. The "7 × 7 reconstruction" specific to Si had previously been observed by electron diffraction measurement, in which Fourier analysis was needed. Binnig and Rohrer were the first to see this famous phenomenon in real space. The periodicity of a crystal such as silicon is interrupted at a surface, a Si vacuum interface. The surface atoms may rearrange to find a lower-energy configuration. So this feature was known experimentally and also understood theoretically, but had never been observed directly, as was accomplished first with the STM.

2. An example of a state of the art cantilever is that in the IBM "Millipede" device. This cantilever is formed as an elongated triangle with the tip mounted at the apex. The length in the direction out to the apex is 50 μm. The supporting arms are 10 μm wide and 1 μm thick and the spacing of the two supporting arms is 20 μm. This device has a resonant frequency of approximately 200 kHz and the effective spring constant is 1 N/m. So 1 nano-Newton would correspond to a nanometer deflection. The tip is extremely sharp, about 20 nm in radius (see Figure 11.1).

3. An electric dipole **p** = e**a** is a vector, a quantity with both size and direction, whose magnitude is the charge times the distance and whose direction is between the centers. Now, an electric dipole creates an external electric field $E(r)$. This field pattern resembles the familiar magnetic field around a bar magnet. The dipole

electric field sticking out from a proton–electron dipole, points out along the radial direction positively away from the positive charge and circles back, like the field lines in a bar magnet. This electric field structure is axially symmetric around the dipole direction. The strength of the electric field characteristically decreases with radius from the dipole, following a law $E \sim 1/r^3$.

4. An attractive force results from the gradient or rate of change of the energy as the two atoms come closer. It is natural to imagine that the two atoms will move together to a place where the interaction energy is more negative. In mathematics, the force is the negative of the rate of change or gradient of the interaction energy. So, $F = -dU/dx$ in a one-dimensional case or $F = -\nabla U$ in the 3-dimensional case. In the case of r^{-6}, the rate of change (derivative) is $-r^{-7}$.

5. The basic law is that, the force is the charge Q times the velocity V times the magnetic field B. In the Hall bar an opposing Hall electric field E_H, leading to the Hall voltage, arises to keep the current flowing in a straight line. The final formula is $E_H = R_H J \times B = (1/ne) J \times B$. And the coefficient of the Hall field, R_H, the Hall coefficient, is found to be inversely proportional to n, the number of electrons/cubic meter in the Hall bar. Using GaAs–GaAlAs conductors the value of E_H (and the corresponding voltage measured in the device) is greatly increased by reducing n and raising J, as seen in this formula.

6. "Heterostructures," alternating slabs of GaAs and GaAlAs, are arranged so that donor impurities in GaAlAs release electrons into the adjacent GaAs layer, where, in addition to the extremely small effective mass, they are free from scattering by the ionized donor ions, which are left behind in the GaAlAs. So smaller electron density n levels and higher current density J are possible when this heterostructure is made into a Hall bar. Both of these effects raise the Hall voltage, and, thus, the sensitivity of the scanning Hall device. The thickness of the GaAs layer is small so that the motion of the carriers perpendicular to the slab is quantized (as in the quantum-dot case we mentioned) while the motion is exceptionally free in the other two directions (in the plane of the GaAs slab). So this is a "2D electron gas" of ideal conducting properties.

7. The superconducting quantum interference detector SQUID in a detection mode rather than a scanning mode is more sensitive, potentially, than the Faraday's law detection that is used in conventional MRI machines. So the SQUID detector has the potential of supporting a much lower-cost MRI technology that would not require the 1-T magnetic field, but rather could operate at a 100-G magnetic field. We will return to this later.

Chapter 10

1. A magnetic moment μ is defined as a current loop I of area $A : \mu = IA$. This is a vector pointing perpendicular to the area A. (A moving charge is the only source of magnetic field B, there is no "point magnetic monopole" that would correspond to a point electric charge, leading to electric field E.) The basic current

loop is an electron in an orbit that has an angular momentum $L = n\hbar$, as we first saw in the Bohr model of the atom. Here, $\hbar$ is $h/2\pi$ and n is any integer. There is a close relation between the angular momentum $L = mvr$ of a mass m moving at speed v in an orbit of radius r, with the formula $\mu = IA$. This is seen by expressing I as e/τ where $\tau = 2\pi r/v$, so that $\mu = (ev/2\pi r)(\pi r^2) = (e/2m)L$. Since the unit of L is $\hbar$, the Bohr magneton is defined as $\mu_B = (e\hbar/2m) = 9.27 \times 10^{-24}$ J/T. If one replaces m (electron) by m (proton) one gets the nuclear magneton, which is 1836 times smaller than the Bohr magneton.

2. The polarization P is defined as $(n\uparrow - n\downarrow)/(n\uparrow + n\downarrow) = (e^x - e^{-x})/(e^x + e^{-x}) = \tanh x$. The value of x is half the energy splitting divided by $k_B T$, $x = g\,\mu_N B/2k_B T$. Because x is so small, the approximation $\tanh(x) \approx x$ applies.

3. Duration, $t_{\pi/2}$, of the Larmor frequency pulse is set by the condition, $t_{\pi/2}(2\pi f_1) = \pi/2$. This was first applied by physicist Isidor Rabi and the frequency f_1 is called the Rabi frequency. It is important in other areas of physics, for example, in applications of pulsed lasers. (Of course this time applies to the application of a magnetic field oscillating at the frequency f_o, which is 42.7 MHz in the case in the text.)

4. We are using the term "Schrodinger cat state" to describe a linear superposition of quantum states, which is not possible in classical physics but is at the heart of quantum physics. Superposition states form covalent bonds (see Chapter 6) and here, in the state of the proton spin, equally up and down, produce precession. In the field of quantum computing, the "Schrodinger cat states" are more specifically described as "equal superpositions of two maximally different quantum states" by Leibfried *et al.*, "Creation of a six atom 'Schrodinger cat' state", Nature 438, 639 (2005).

5. E.W. spent 2 years at the University of Illinois, where John Bardeen was Prof. of Electrical Engineering and Prof. of Physics. While at Illinois, EW discovered an interesting effect using the (then) new technique of electron tunneling to study electron–phonon interactions in p-type Si. John Bardeen helped in the explanation of the interesting effect, and Bardeen's help was acknowledged by EW in the resulting paper: E. L. Wolf, Phys. Rev. Lett. 20, 204 (1968).

Chapter 11

1. To make a suspended Si spring layer, as is needed in the IBM "Millipede" device, one needs a "buried layer" or *sacrificial layer*, typically SiO_2. This can be done by thermally oxidizing a Si surface, and then growing a layer of Si onto that layer, subsequently annealing the structure to allow a clean demarcation of Si and its oxide. Alternatively, one can heavily ion bombard the Si surface with oxygen ions at an energy to penetrate deeply into the Si. The heavily oxygen-implanted Si can be annealed (heated) and under the right conditions an epitaxial Si-SiO_2-Si structure can be achieved. (A buried oxide layer "BOX" is sometimes used as a starting substrate to make FET devices. One can selectively remove SiO_2 with HF

(hydrofluoric acid), which does not etch Si itself.) The BOX layer in the "Millipede" is evident in Figure 11.2, at the left side, where the cantilever is attached to the underlying Si.

2 A few more aspects of the "Millipede" system are worth mentioning. The speed of reading and writing is partially dependent upon the thermal time constant of the tip. The tip thermal time constant controls heating and cooling times for the tips. Because the tips are so small, these times are on the scale of microseconds. A µs is probably long enough to read the resistance of the tip heater which is the means of detecting a dimple (vs. no dimple, the binary information of the system). One would guess that the dimple writing time would also be on the order of µs. The polymer itself is likely to soften very quickly as the tip is heated and to allow the motion of the tip into a surface. The vibration frequency of the tip stated to be 200 kHz would correspond to a period of 5 µs. So perhaps this would be the limiting feature in the time to write a dimple as the tips are raised and lowered with respect to the surface, one might expect vibration of the tip to persist over a number of cycles. The Q of the tip oscillation is probably high and so the time for the tip to stop oscillating after being moved up and down with the silicon surface might be a multiple of those periods of that oscillation, which are 5 µs. So this effect, unless a damping is provided in some way, would result in a slower writing and reading time for this system.

Another aspect is temperature control. The array of 1024 tips basically aligns with respect to the data set underneath because both are mounted on single crystals of identical materials. So thermal expansion of the crystal holding the cantilevers and of the crystal holding the dimples, should be closely similar. To achieve the data accuracy of 10 nm, it was found necessary to temperature control the two chips of silicon to 1 °C, which was done by heaters placed on the corners of the Si chips (Figure 11.3). The cantilever array, as we said, is on a Si chip whose outer dimensions are of a 7×14 mm size. The high-frequency 200 kHz of the tips suggest the danger of acceleration would be minimized. On the other hand, the relatively massive mm-sized chips have to be stable against mechanical shock and this will appear to be a problem. The vibration frequency of one Si chip with respect to the other is certainly much slower than the 200 kHz motion for the tip. It is possible that the interchip motion may have been damped to minimize the sensitivity of the system to external shock leading to relative acceleration between the cantilever chip and the data chip.

3 The identical nature of electrons, and the rule of nature that says that the wavefunction for two electrons must *reverse sign* when the *particles are exchanged* are the basis for the covalent bond and also for ferromagnetism. Ferromagnetism is at the heart of the 1-TB hard drive that is part of the information explosion. The topic also relates to the structure of matter via the Pauli exclusion principle, which says that there are only two electrons (spin up and spin down) allowed per quantum state. Here we see that the rules built into our Universe, beyond the existence of Schrodinger's equation, are needed to understand important behaviors. The topic of particle exchange is simple but can be confusing, because we

have to extend to the case of a *two-particle wavefunction* the ideas of linear combinations of states, with the idea of distinct wavefunctions for particle *locations* and particle *spin*, with the symmetry requirement that the complete wavefunction must reverse sign on particle exchange.

4 If we take the two protons of hydrogen, and put in the two electrons, the physics has to add up the energies that are attained using the various wavefunctions. It is found by mathematics based on Schrodinger's equation, that for a hydrogen molecule, the most stable state is the spin singlet. This is space symmetric for the reason just stated, and the electrostatic energy is more favorable. The binding energy is 4.5 eV, which is substantial, and the difference to the spin triplet state, (antibonding state) is about 9 eV. The two states are split almost symmetrically about zero. And recall that the antibonding state, having the antisymmetric space function, with a node at the center, does not allow as much attraction for electrons to the two positive protons. The calculation can be done at any spacing between the protons, to find the most stable spacing of the atoms. The energy is most negative at 0.074 nm, the bond length in H_2. (The same arguments apply for the directed wavefunctions in Figure 6.5, these are two-electron covalent bonds.) However, the ordering of the singlet and triplet states reverses for more complicated wavefunctions, such as those with quantum numbers $n = 3$ and $l = 2$ (so-called "3d") in Fe and Ni. If the same Schrodinger analysis is applied to Ni and Fe, stability is obtained for the *spin triplet state, parallel spins*. The electrostatics, through the symmetry requirement, forces these spins parallel, giving ferromagnetism.

Chapter 12

1 A capacitor C stores charge $Q = CV$ at applied voltage V. The electrostatic energy $U = Q^2/2C = CV^2/2$. For very small capacitance values the energy change $\Delta U = e^2/2C$ in changing the charge by one electron charge can become important, which means it can become comparable to or larger than the thermal energy, $k_B T$. Typically the capacitance decreases with device size, actually the capacitance of an isolated sphere of metal of radius a is just $C = 4\pi\varepsilon_0 a$. The "Coulomb blockade" condition $e^2/2C \gg k_B T$ generally requires an atomic size capacitor or a cryogenic temperature. For this reason single-electron transistors are not available for operation at room temperature.

2 An attofarad (aF) = 10^{-18} F. The other common use of "a" = atto is in description of times, an attosecond is 10^{-18} s. To make a rhyme (or a difficult crossword puzzle clue), an atto is a nano-nano? In a recent announcement of the "L'Oreal-UNESCO Awards 2011 for Women in Science" (New York Times, Tuesday, March 8, 2011 p. D8), the list included Pr. Anne l'Huillier of Sweden, whose field of research was identified at the bottom of the page as *"Attophysics."* It turns out that Prof. l'Huillier is an expert in the use of short laser pulses, on the *attosecond* time scale. We had never before encountered the term "Attophysics." However, to avoid a

proliferation, which would also include "picophysics" and "femtophysics," we suggest that "nanophysics" can serve for nonexperts to denote physics on size scales *smaller than* 100 nm. (This makes sense because it is the single boundary between classical and quantum physics, everything below 1 nm follows quantum rules.) Hence, "attophysics," which we take to mean physics on the size scale of 10^{-18} m, would formally be included in "nanophysics." Nuclei are objects on the femtometer scale, 10^4 or so times larger than 10^{-18} m. The smallest meaningful size in the physical world is believed to be the Planck Length 1.616×10^{-35} m, associated with the Big Bang origin of the Universe.

3 The capacitance of the nanotube can be estimated, for a length L from the classical formula $C = (2\pi\varepsilon\varepsilon_o L)/\ln(b/a) = (2\varepsilon L/k_c)/\ln(b/a)$, where b and a, respectively, are the outer and inner diameters of the assumed cylindrical capacitor. This formula is useful even if the geometry is not perfectly cylindrical, and the parameter b is taken as the distance of the nanotube through the Si oxide to the underlying Si chip.

4 The radio-frequency single-electron transistor RFSET device is a lithographically constructed SET, with dimensions on a micrometer scale, using electron-beam lithography to achieve small capacitance tunnel junctions between the island and the source and drain (Schoelkopf *et al.*, Science 280, 1238 [1998]). Nevertheless, the resulting capacitances are too large for room-temperature operation, and this device is operated at cryogenic temperature. A novel approach to readout makes it a practical and useful electrometer. The novel readout avoids high operating resistance levels that would otherwise lead to slow response times. The RFSET readout scheme is similar to the conventional detection of electron spin resonance (ESR), using the change of reflection of (high-frequency) microwaves from a high-Q resonant circuit (cavity) when the cavity experiences internal losses (damping).

The idea is to use "RF, radio frequency" to monitor the *drain–source impedance*, which quickly takes different values as the charge on the island (between the source and the drain) goes from half-integral (low impedance) to integer values (high impedance, due to Coulomb blockade). This quickly varying *drain–source impedance (which contains information about the island charge)* is made part of a series-resonant high-Q circuit, driven by the 1.7-GHz "carrier" fed in a coaxial line. The reflection of the microwaves is detected, which changes rapidly with change in impedance of the source to drain connection as the island occupation changes. This technique allows a high-frequency readout of the SET island occupation.

This device accurately measures a 30-kHz triangle wave $Q(t)$ about 5.5 electrons on an island. It is seen that the RFSET response easily and quickly records the island (gate) charge in units of e, with very small noise. The gain of this device is nearly constant up to 10 MHz.

5 The nanotube is classified by integers (n,m) and the diameter is $d = a_H (n^2 + m^2 + nm)^{1/2}$ where $a_H = 0.249$ nm is the lattice constant of the honeycomb

carbon lattice of the graphene layer (one layer of graphite). The case $n = m$ means that the tube is achieved by rolling the hexagonal lattice around an axis through the apices of its hexagons, leading to the "armchair" appearance of the atomic arrangement. Armchair tubes are metallic. Tubes in the case $m = 0$ are called "zigzag" present a sawtooth appearance, and these are metallic only when n is a multiple of 3. The diameter of the (10, 10) metallic nanotube is thus 1.36 nm. Tubes intermediate between these two cases, such as the STM image in Figure 7.2 are called "chiral."

6 In the formula[3] taking values $L = 1\,\mu m$, $a = 60\,nm$, $b = 140\,nm$, and $\varepsilon = 7.5$ (for Si_3N_4) gives

$$C = (2 \times 7.5 \times 10^{-6})/9 \times 10^9)/\ln(140/60)$$
$$= (1.666 \times 10^{-15})/\ln(2.333) = 1.41 \times 10^{-15}\ F = 1.41\ fF$$

7 The orientation of the 16 spectator molecules is vertical, and the oxygen atoms are arranged radially. The central flat molecule immediately face 16 oxygen atoms at the inner apex of the 16 vertically oriented benzene rings. The methyl groups have a rotational degree of freedom, but only the outer methyl groups on the 16 spectator molecules are in fact capable of being reoriented. The individual molecules are methyl-substituted 1,4 benzoquinone (DRQ), in full chemistry notation this is "2,3,5,6-tetramethyl-1-4 benzoquinone." These structures were formed by depositing the DRQ molecules on a gold surface and manipulating them with an STM tip. It was found that adding one more DRQ by STM tip nudging into 16 membered semicircular flat array leads to the closed assembly, 16 (plus one) where the 16 have now jumped into a vertical orientation.

Chapter 13

1 These intractable calculations have been commonly used as encryptions to safeguard bank accounts and other valuable information. A breakthrough in quantum computing would have wide consequences in areas including banking and finance. So while quantum computing, on the one hand, offers potential efficiency, on the other hand, it would present quite a disruption to the existing financial industry.

2 This $D-D^+$ system is analogous to the H_2^+ hydrogen molecular ion known to chemists as an example of the "one-electron bond" and the one-dimensional analog was shown in Figure 6.4. The D_2^+ donor case differs, in that the antisymmetric states (which do not exist for H_2^+ in vacuum), are bound for D_2^+ in silicon, thus available for use in the computational schemes. Although in the vacuum version of the hydrogen molecule ion the fully entangled antibonding state does not exist, it does exist in replica molecules formed using donor states (Chapter 7) in semiconductors. The "buried-charge qubit" is an example of a two-level system, similar to a spin ½. All of the manipulations that were described in

connection with spin ½ in the MRI apparatus have analogs in the charge-qubit case.

3 In the case for a time-varying interaction, where H is the sum of the kinetic and potential energies as in Chapter 6, the Schrodinger equation is $\hat{H}\Psi = E\,\Psi = i\hbar\partial\Psi/\partial t$. In the context of data handling, the wavefunction Ψ is a data vector, and an initial vector is chosen that corresponds to a known interaction H. The idea of the adiabatic quantum computing is to slowly change the H to match the statement of the problem, and to follow the evolution of the data vector, which according to the Schrodinger equation and the idea that for slow variations the wavefunction will adjust to follow the changes, and in its final state will represent the answer to the problem. If we have 32 electron spins, we might start them all pointing to the positive x-direction, and we know exactly how that is a solution to the starting H function.

In simple terms this is like rotating a vector, for example, the minute hand on a clock. If we consider xy coordinates the point 0,1 (12 on the clock) can be transformed into the state 1,0 (3 on the clock) by rotating the initial vector by $\pi/2$ clockwise. In the analogy 0,1 would be the initial wavefunction vector 1,0 is the answer and its rate of change is provided by the rotation operation, in the example clockwise by $\pi/2$, which is provided by the H operator in the quantum case. In physics, the H operator acting on the wavefunction vector rotates that vector. In the real case for a 32-qubit quantum computer the vector would point in a space of 32 dimensions, it would start from a solution vector to a known problem and the H operator, into which the problem is coded, would slowly rotate the vector to its final orientation, representing the answer to the calculation, which is another vector in 32 dimensions.

4 A *Josephson tunnel junction* is a tunnel junction between two superconductors, which has a sufficiently thin barrier that the superconducting pair wavefunctions on the opposite sides significantly overlap, creating energy E_J of coupling between the two superconductors across the junction. If this energy is larger than $k_B T$ it allows a *Josephson supercurrent* to flow across the junction. The size of the supercurrent I_J depends on the difference $\Theta_1 - \Theta_2 = \varphi$ of the superconducting phases on the opposite sides as $I_J = I_{Jo}\sin\varphi$. These devices are also used in SQUID detectors mentioned in Chapter 10.

5 The voltage pulse as a flux quantum escapes from the loop containing two Josephson junctions can be described by integrating Faraday's law over time, to get $\int V dt = 2.07 \times 10^{-15}\,\text{W} = 2.07\,\text{mV ps}$.

Chapter 14

1 Claytronics, a concept developed in collaboration of scientists at Carnegie Mellon University and Intel Corporation, is a form of programmable matter made out of very large numbers, potentially millions, of submillimeter-sized spherical robots. These work together to form dynamic 3D physical objects. For example,

claytronics might be used to mimic the actual presence of a person, as in a telephone call (S. C. Goldstein, *et al.*, "Beyond Audio and Video: Using Claytronics to Enable Pario" AI Magazine, 30[2] July 2009).
2 Science May 2010 Gibson *et al.* "Creation of a bacterial cell controlled by a chemically synthesized genome".

Index

a

Abacus   17, 151, 179
Accelerometer, using cantilever   24
Actin, in molecular motor   41, 43
ADP and ATP, in molecular motors   42
Artificial intelligence (AI)   9
– Practical form of   9
– "Singularity" in   9, 166, 170–171, 174–175
– Strong form of   10
Artificial life (bacterium)   169–170
Atanasoff, John Vincent
– As inventor of digital computer   179
Atom   1, 18, 57
– Binding energy of electron in   59
– Emission of light by   64
– Nucleus of   7, 27
– Size scale of   18, 57, 59
– Wavefunctions for   73, 76
Atomic force microscope (AFM)   110–114, 127
– Sensing by light deflection   110–111
– Sensing by van der Waals force   111–114
– Use to form SET in carbon nanotube   143
Atomic layer deposition   103

b

Bacterium   43, 56
– Artificial   169–170
– As scaling limit of submarine   55–57
– Complexity of   183–184
– DNA replication in   49
– Forming nanotube   88
– Magnetic   56
Bardeen, John   122–123, 191
Benzene rings to form molecular computer   147–148, 195
Brownian motion, as evidence for lumpiness of matter   57
Buckyball C_{60} 91, 92
– In single-electron transistor   140
Bumblebee, flight of   39

c

C_{60} see buckyball
Cantilever   31, 127, 129, 181, 189
   In accelerometer   24
Carbon atoms   76
Carbon nanotube see nanotube
Cesium   5, 7, 14, 64–5, 75, 77, 184
Clocks   4, 14
– Atomic   5, 27, 184 see also cesium
– Harrison chronometer   4, 181
– In PC computer   30, 34, 182

Understanding the Nanotechnology Revolution, First Edition. Edward L. Wolf, Manasa Medikonda.
© 2012 Wiley-VCH Verlag GmbH & Co. KGaA. Published 2012 by Wiley-VCH Verlag GmbH & Co. KGaA.

- Molecular (in genomics) 50
- Pendulum (grandfather) 31, 36
- Quartz 5, 31, 34
- Spring 31, 181

Cloud computing 21

Covalent bond
- Energy of 81
- From directed wavefunctions 74
- In double-well potential 71–72
- Physics of 132–134, 192–193
- Probability of breaking 8

d

Digitized books 163–164
Dimples, as data points 131
Disk *see* magnetic disk data storage
DNA 2, 46
- Information capacity of 48–49, 183
- In nanofabrication 91
- Replication of strands 46
- Sequencing of 47, 50–51

e

Einstein, A. 14, 26, 57, 100
Electron 58
- Spin of *see* spin

Energy availability
- Cost of computing 152, 160
- Energy bubble in era of cheap oil 171

Exclusion principle for electrons 76
- And chemical table 76–77

f

Facebook 13, 21, 170
Ferromagnetism, ferromagnets
- Effect on electron bands in metal 135
- In magnetic disk data storage 132, 181
- Nanometer size, in bacteria 52, 56
- Physical origin of 132–134, 192–193
- Soft and hard 24

FET *see* transistor
Fission, nuclear 7
Frequency
- Of counter 27
- Quartz oscillator (PC clock) 34, 182
- Quartz tuning fork (cantilever bar) 31, 181
- Resonant 24, 129, 181
- Scaling of, for cantilever bar, dimensional and materials aspects 31, 181
- Scaling with size of 36, 181

g

Genome (human) *see* technology, genomic
Genographic Project 2 *see* technology, genomic
Global positioning system (GPS) 6

h

Hard disk 102, 181 *see also* magnetic disk data storage
Homo sapiens 1, 6, 177
- Migrations of 2, 50, 177
Human progress 1, 17
Hybrid technology 7, 14
Hydrogen
- Atom, Bohr's model of 57, 59
- Atom, light emission by 64, 184
- Molecule 36, 132–134, 192–193
- Wavefunction for 73

i

Information technology 6, 17, 178
- Exponential growth of 171, 173–174, 178
- Moore's law, in semiconductor devices of 6, 13, 19, 22, 27
- Revolution in performance of 178
- Role of nanotechnology in 178
- Role of search engines and digitization in 163–164, 170

j

Josephson tunnel junction   124, 158, 196

k

Kaku, Michio (Physics of the Future)   15

Kurzweil, R. *see* Singularity, artificial intelligence

l

Laser (light amplification by stimulated emission of radiation)   14, 100, 181
– Excimer ArF laser at   193 nm 100–101
– In P-N junction injection laser   26, 98–100

Light
– As photons, in technology   64
– Spectrum of, from hot object   57, 58

m

Magnetic disk data storage   17, 102, 127, 181
– Reading device   17, 22, 127

Magnetic force microscope   113

Magnetic moment *see* spin

Magnetic polarization, P   191

Magnetic resonance imaging (MRI)   14, 65, 117–126
– Faraday's law detection of rotating pixel moments   120
– Larmor precession: 42.7 MHz/Tesla 118–120
– Pixel (1 mg of H_2O) and its scanning 118–121
– Relation to spin ½ for proton 117–119
– Resonance, Rabi frequency and Schrodinger's cat   120–121
– SQUID detection scheme for MRI 122–125

Magnetic tunnel junction (MTJ)   17, 127, 180
– As magnetic disk reader   22, 132, 181
– Spin-specific tunneling in   26, 132, 180
– To determine spin orientation   26

Mesa, etched, in silicon   128

Metal as electrons in empty box (3D trap)   77–78
– Fermi energy in   78, 188
– Metallic bond, cohesive energy   78

Millipede, data storage device   127–131, 168, 192

Molecular motors in biology   41–45
– Linear   41
– Rotary   43
– Torque and power in   44–5, 182–183

Molecule   108
– Acetylene, on STM tip   109
– ArF (excimer)   100
– Biphenyl, formation of, by STM   108
– C_{60} 86, 91–92
– Double-well model for   71
– H_2O as detected in magnetic resonance imaging (MRI)   119, 126
– $HfCl_4$ 103
– Hydrogen   36, 132–134
– Methane   74
– Polar   89–90
– SiF_4 88
– Vibration of, measured by STM   109

"Molecular assembler"   166–168
  Connection with Millipede device 168
– Refutation of concept   167–168

Moore's law *see* information technology

MTJ *see* magnetic tunnel junction

Myosin, in molecular motor   41

n

Nanometer   18–19, 29

Nanophysics   7, 14–15, 17, 22, 24, 26–27, 67–72, 181, 193–194
– Definition of   7
– Technologies based on   26–27

Index | 201

Nanotechnology   1, 5, 9–15, 17, 181
– Broadened definition of   14
Nanotube, carbon   31, 86, 187
– Electrical field of   89
– Growth of   87
– In Damascus blades   90, 187
– Radius of   89, 194
– Use as detector of polar molecules   89
– Use in dense nonvolatile memory array   143–147
– Use in single-electron transistor (SET)   142–143
Nucleus (of atom)   7, 178
– Carbon   8, 76
– Cesium   5, 7, 14, 64–5, 75, 77, 184
– Fission of   7
– Strong force in   75
– Uranium   7
Neutrino   185
Neutron   75, 178, 185

o

Optical fiber   5, 6, 8, 14, 26, 189
– Combined with injection laser   98
– Single mode   26
Oscillator
– Bulova quartz   33
– Molecule vibration   36
– Pendulum   31
– Spring   31–32, 181
– Tuning fork   31

p

Paleomagnetism   183
Photolithography, patterning chips   128
– Photoresist   128
Photon   59
– Generation of injection laser   98–100
– Roles in technology   64
Picoliter reactors, in DNA sequencing   50
Planck's constant   58, 184
– Role in averting atomic and cosmic collapse   60
Planck length   179, 194
P-N junction, in semiconductor   93, 98, 185, 189
– As injection laser   26, 98–100
Pollen particle   38
Proton   58
P-N junction   96, 98–100

q

Quantum computing   152–159
– Adiabatic form   157, 196
– Qubits in   152–153, 154–157
– Searching capability   153
Quantum dot   29, 30, 85–86
Quantum physics see nanophysics

r

Rapid single flux quantum (RSFQ) computing see superconducting computing

s

Scaling the size   27
– End of   63
– Frequency of cantilever bar   31, 189
– Of Captain Nemo's submarine   55–57
– One-dimensional vs. three-dimensional (isotropic)   29, 36
– Power density change   37
– Viscous force change   39, 44
Scanning Hall microscope   114
Scanning near-field optical microscope (SNOM)   115
Scanning SQUID microscope (for magnetic flux)   115, 124
Scanning tunneling microscope (STM)   105–108
– Piezoelectric positioner in   105–106
– Scanning tip in   106–107, 109
– Single atom or molecule tip decoration   107–108

– Use for nanofabrication 107–109
– Vibrational frequency determination by 109
Schrodinger's cat states 121–122, 165, 186, 191
Schrodinger's equation 67
– For double-well potential 71, 72
– For linearity 165, 186
– For trapped particle 69–73, 77
– Probability distribution and wavefunction resulting 69
– Tunneling effect 68, 69, 72, 180, 185
SET *see* single-electron transistor
Shoebox, etched mesa, in silicon 128
Silicon, a prototype semiconductor 81 ff, 93 ff, 188
– Bands for electron conduction in 83, 93–94
– Covalent bonding in 74, 81
– Electrical resistance of 83
– Electrical resistance when chemically doped 84, 93–94
– Electron concentration, when pure 82, 188
– Nanowires of 88
– P-N junction in 95, 111
Single-electron transistor (SET) 139–143
– Based on carbon nanotube 142–143
– Radio frequency version, RFSET 194
"Singularity" *see* artificial intelligence
Spin—angular momentum and magnetic moment 165, 184, 191
– In technology 65, 117–126, 165
– Of electron 132
– Of proton 58, 118
– Polarization (magnetic), P 191
– Singlet and triplet combinations of two 133
Spring constant K (Newtons per meter) 32, 36
– Resonant frequency 37

Superconducting quantum interference detector (SQUID) 123–125
– Josephson tunnel junctions in 124, 196
– Quantized flux 125
Star-stuff 185
Superconducting computing: rapid single flux quantum (RSFQ) technology 158–160
– Josephson junctions in 158–159, 196
– Tipping point for large server installations 160
– Use for analog to digital converter (ADC) 159–160
Superconductivity
– Energy efficiency of 160
– Flux quantum 124
– Possibility of room temperature form of 173
– Quantized flux in loop 123–124

t
Technology
– Acceleration of 1, 8
– Cascade of several 6
– Clocks 4
– Elements of, based on nanophysics 26
– Genomic 2, 5, 16, 46 ff, 197
– History of 14, 26
– Hybrid forms of 7, 14
– Industrial revolution in 2
– Information 6, 17 *see also* information technology
– Nanometer scale *see* nanotechnology
– Relation to human progress 1, 17
– Synthesis in 2
Terabyte (TB, 10^{12} bytes) 22, 127, 179, 183, 192
Time, short history of 3
Transistor 6, 19

– Field effect transistor (FET)   29, 85, 101–04
– High kappa oxide as gate insulator in FET   103
– P-N junctions in   95, 111
– Scaling law of FET   39, 102–03
Tunneling magnetoresistance (TMR)   180, 181
– As measured by magnetic tunnel junction (MTJ)   23
– In magnetic disk reader   23

u

Uncertainty principle   185

v

Van der Waals force   111–114
– Role in atomic force microscope (AFM)   112
Venter, Craig   169

w

Watson (IBM computer system)   6
– As example of practical artificial intelligence (AI)   6, 12
– Winner of Jeopardy   12

Wavefunctions of electrons
– As predictors of particle position   69
– Delocalization and kinetic energy   78–79
– For spherical atom   73
– For trapped particle   70, 78
– Linear combinations (hybrids) of   72, 165
Waves
– DeBroglie length h/p of   62
– Diffraction of, for light and electrons   62
– For matter, and Schrodinger's equation   63, 67
– Relation to particles of nature   60–63, 67
– Speed of, related to length and frequency   61
– Tutorial discussion of   60–61
– Wavefunction and probability   63
Whales and elephants, motion of   41, 45

y

Young's modulus, shear modulus   30–31